Forecast is the most urgently needed [...] in a very long time.

Joe Shute is one of Britain's finest [...] other subject.

John Lewis-Stempel

This urgent, elegiac book's call to mend our broken relationship with the land feels more vital by the day.

Mail on Sunday

With a journalist's eye for detail, Joe backs up his captivating anecdotal evidence regarding the seasons with the results of solid scientific research to finger the culprit: global warming.

BBC Countryfile

An absolutely beautiful account of life going on while the world stopped. I loved it.

Kate Bradbury

Joe Shute illuminates in beautifully clear prose, laced with well-judged literary and historical references, the scale of the threat posed to our natural world by Climate Change. A 'must read' for anyone who is curious and who cares.

Jonathan Dimbleby

This is no ordinary nature diary – it enlarges our perspective of what has altered, and what is being lost [...] one of the most poignant and affecting nature books I have read this year.

Miriam Darlington

What a wonderful read. Told through the eyes of farmers, poets and philosophers as well as the author's own personal explorations across the country, *Forecast* is a beautifully written elegy to our natural world and a warning of how quickly it is changing.

William Sieghart

Forecast is a triumph of the most unnerving sort.

Simon Ings, *The Daily Telegraph*

Full of information and very enjoyable.

Bird Watching

A Note on the Author

Joe Shute is an author, journalist and weather watcher with a passion for the natural world. He writes features for *The Daily Telegraph* and is the newspaper's long-standing Saturday 'Weather Watch' columnist. He is currently a postgraduate researcher funded by the Leverhulme Unit for the Design of Cities of the Future at Manchester Metropolitan University.

Joe studied history at Leeds University and started his career as a trainee reporter on the *Halifax Evening Courier* before working at *The Yorkshire Post* as its crime correspondent. He previously wrote *A Shadow Above: The Fall and Rise of the Raven*, published by Bloomsbury in 2018. He lives with his wife in Sheffield.

@JoeShute

FORECAST

A Diary of the Lost Seasons

Joe Shute

BLOOMSBURY WILDLIFE
LONDON · OXFORD · NEW YORK · NEW DELHI · SYDNEY

BLOOMSBURY WILDLIFE
Bloomsbury Publishing Plc
50 Bedford Square, London, WC1B 3DP, UK
29 Earlsfort Terrace, Dublin 2, Ireland

BLOOMSBURY, BLOOMSBURY WILDLIFE and the Diana logo are
trademarks of Bloomsbury Publishing Plc

First published in the United Kingdom 2021

A catalogue record for this book is available from the British Library

ISBN: HB: 978-1-4729-7674-1; TPB: 978-1-4729-7677-2;
eBook: 978-1-4729-7676-5

2 4 6 8 10 9 7 5 3 1

Typeset in Bembo Std by Deanta Global Publishing Services, Chennai, India
Printed and bound in Great Britain by CPI Group (UK) Ltd, Croydon CR0 4YY

To find out more about our authors and books visit www.bloomsbury.com
and sign up for our newsletters

For B

Contents

So long as men can breathe or eyes can see
So long lives this, and this gives life to thee

Shakespeare, Sonnet XVIII

The seasons are against us.

Professor Chris Whitty, Chief Medical Officer
for England, 21 September 2020

CHAPTER ONE

A Lockdown Spring

It takes three weeks for spring to travel up the country, moving north-east at a rate of about 2mph. As I drove along the A1 in the early days of the first spring of a new decade, I wondered at which point I had passed it by. There was nothing else to overtake but the rising sap, roadside blossom and unfurling leaves. The Great North Road was so deserted I could drift across the white motorway lines.

Buzzards starved of roadkill perched on the treetops staring blankly over the empty asphalt. Some deer had ventured out of a patch of trees to graze the grass verges, nitrogen-enriched from the exhaust pipes of the tens of thousands of vehicles that on any usual day would be passing by. Insects spattered against the windscreen with a ferocity I had not seen since childhood, smearing a blood mosaic across the glass.

The road signs blinked *STAY AT HOME*. Police patrols were randomly stopping the few cars they could spot. But in my pocket was a piece of paper granting me special access to this spectral road. In the early days of the 2020 lockdown, brought in to contain the spread of the Covid-19 pandemic, I had suddenly gone from being a journalist to a 'key worker'. And my newspaper had tasked me with travelling the length and breadth of the country, from north to south and coast to coast, to discover how communities were responding to the deadly new virus in our midst.

Over the course of that week on the road, my photographer colleague and I stopped at motorway service stations populated solely by the animated screens of Costa Coffee self-service machines. We passed Stonehenge, where the empty car park – normally filled with coaches – reminded me of a go-kart track. At night we stayed in any hotel we could find that had remained specially open for key workers.

One dilapidated seafront guesthouse near Eastbourne we shared with doctors working on the Covid wards, who didn't want to go home for fear of spreading the virus. As we passed one another on corridors that smelt of hand-sanitiser and bleach, we each shrank into our side of the wall, doing our best to crease our masked features into a recognisable smile.

Another hotel, an imposing Victorian building with sweeping views across Morecambe Bay, reminded me of the Overlook in *The Shining*. We were the only two guests staying among hundreds of empty rooms spread over two wings. The cavernous dining room was perfectly laid for a breakfast that never happened. We took our meals as instructed, alone in our rooms, watching the news for updates of the daily death toll and of the condition of the Prime Minister, who was in intensive care after contracting the virus.

Do you know what the people we met wanted to speak most about that week, as life as we knew it transformed before our eyes? The weather. That lockdown spring was, paradoxically, the brightest on record. March gave way to the sunniest April, which in turn became the sunniest May ever witnessed (at least since records began in 1881). It was the fourth driest May, too. As every day dawned to another endless blue, I had a sense of Covid-19 bending time.

So many of the seasons which normally dictate our year were, in a matter of days, rendered meaningless. The football season, the fashion season, the fishing season, the wedding season, academic calendars and holiday dates, all evaporated from diary pages as if drawn in invisible ink.

Four seasons, however, remained. Their passing suddenly gained an importance that had been forgotten in the hurried turmoil of modern lives. People spoke of the scales falling from their eyes and wondered if the birds always sang this loud, or had we just started listening to them? Amid the spiralling death toll and fear of every interaction with a stranger, spring, the season of regeneration, erupted in glorious technicolour.

My lockdown road trip ended in the Derbyshire village of Eyam, where in 1665 residents cut themselves off from the outside world for eighteen months after an outbreak of bubonic plague, to prevent it spreading to the surrounding area. I interviewed residents who in some cases could trace their families back to the original epidemic and were now shielding from coronavirus in the very same seventeenth-century cottages as their ancestors once had. Standing there talking to people the agreed distance of two metres away in the threshold of the old plague cottages, rarely had I been so aware of the elasticity of our ties to the past.

I returned home to nearby Sheffield to commence my own period of self-isolation. Among the strange must-haves during that period – sourdough starters, tomato plants and so on – was a copy of *The Plague* by Albert Camus. My mum sent me one in an early Covid care parcel and reading of the fate of the people of the Algerian port city of Oran, gripped by an outbreak of bubonic plague, I was struck by similarities with our own situation. Not least, the weather. Camus wrote of the 'incessant sunshine' weighing heavily on the plague-ridden town as the death toll crept up. The French word for weather, *temps*, is the same as for time, and I started to feel a strange interlocking between the two.

Weeks blended into months as I watched spring pass. Every morning before settling down to write, I took a walk through my local woods. In my garden I counted the species of birds and insects and the flowers that had come into bloom. This in itself is nothing particularly new. Since moving here with my wife three years previously, we had both always kept a close eye on life in our garden. But suddenly I was able to watch the season on a level of intimacy my peripatetic life as a journalist never normally affords.

I found a baby newt in the pond on 16 March, heard the first chiffchaff of the year on 18 March and on 24 March spotted a tortoiseshell butterfly dancing across the forget-me-not sprays. I watched crows digging up moss from the lawn to soften the nest they had built in an ash tree at the end of our garden and a pair of goldcrest excavating the old brick garden wall for the same purpose. A pair of dunnocks built a nest in a climbing hydrangea on the side of our house. One morning I found a tiny blue egg that had toppled out and cracked on the paving slabs below.

On 18 April came a reminder that it is not just viruses that speed across the world. 'Hirundo Domestica!!!' wrote eighteenth-century naturalist Reverend Gilbert White in his diaries, marking the first swallow of the year. I spotted my first swallow of 2020 as I cycled to the top of a hill called

Long Line in Sheffield during my hour of daily exercise; it swooped low over my head, its fork-shaped tail streaming behind.

I had come here hoping to see swallows. The road, which as its name suggests heads dead straight from the city's southern suburbs up to the Peak District, often swirls with them in the summer months, though I had never spotted them quite so early before.

I wrote an article about my swallow sighting in the newspaper and received a flurry of responses from readers who had just seen their own first birds of the year. Beaminster, Dorset on 9 April, the Lizard Peninsula on Easter Saturday, Oakington near Cambridge on 17 April, and on the twenty-first a solitary swallow over Blenheim Palace. The messages contained little more than this basic information but each one told me a story of quiet satisfaction: a sense of surety that even in the grip of a pandemic the swallows still come back. The poet Ted Hughes's oft-quoted line about swifts is equally applicable to their hirundine cousins, the swallow. Their return marks a sense that the world is still working.

Another reader got in touch, a man called Graham White (no relation, I presume, to the aforementioned Reverend Gilbert), who mentioned to me he had seen his first swallow that year over the Cotswolds on 5 April. Graham told me he had been keeping the dates of swallow arrivals since 1977. Back then the birds would first arrive at the very end of April, but slowly as the decades have passed the date has crept forward. The previous year, Graham wrote, the first birds were spotted on 30 March.

He admitted that rereading his old swallow lists stirred up fragments of memory: of former girlfriends, and of once thumping a man for cheating at a fishing contest. These are the links between landscape, weather, memory and nostalgia that I want to explore in this book: the hawking swallows just one of many needles stitching together the tapestry of the year. How the weather, and the great seasonal orchestral moves it conducts, shapes our own sense of time.

I also want to explore how this relationship is shifting, profoundly so, in the era of climate change. To bridge the void between our cultural expectation of the seasons and what they are actually doing. To follow the march of the seasons up and down the country and document how their changing patterns affect all of our lives. And to discover what happens to centuries of folklore, identity and memory when the very thing they subsist on is changing, perhaps for good.

Back in the eighteenth century, for example, the first swallows would sometimes not be sighted until June. White, along with other naturalists of the period, felt the idea of migration was a flight of fancy. More likely, they presumed, the birds were hibernating underwater during winter, or hunkering down in chimney pots to see out the cold months in a state of torpor.

As we have unpicked the secrets of their sub-Saharan migration in recent decades, that epic journey has started to alter as the birds, like us, try to make new sense of the changing seasons. Studies by the likes of the British Trust for Ornithology have found that swallows are these days arriving from their African migration a fortnight earlier than in the 1960s, and breeding a full eleven days earlier. The warming climate is changing what we presumed to be fundamental seasonal rhythms.

In recent years, swallows have been spotted by English county recorders as early as February, while there have been numerous reports of the birds even over-wintering in the south of the country. In doing so they are answering the prayers of the pastoralist poet John Clare who longed for swallows to brighten up the darkest days, 'twittering as wont above the old fireside/ and cheat the surly winter into spring'.

In late May I received another snippet of swallow news – and it wasn't good. A man called Chris Jones, who runs an organic cattle farm in mid-Cornwall, which has been in his family since 1960, spanning three generations, had for the first time in his life not seen any swallows arrive. Usually they stream in across the 170 acres he farms in early April, building their nests, like earthen hanging baskets, in the corners of his old stone barns and swooping low over the meadows in pursuit of the insects rising up from the steaming cattle dung. On any given year he receives about thirty nesting pairs, but during 2020 – save a few lone birds overhead – not a single one came to stay.

When we speak by phone it is a windy day and Chris, who is sixty, says normally he would delight in watching the birds take a breather on the telephone wires strung across his land. By then the drought that had accompanied the record sunshine had started to take hold and his pastures were beginning to brown over. Years ago he lost his breeding pairs of cuckoos, another harbinger of early summer with deep cultural roots, but the absence of the swallows hit him particularly hard. As long as he has been here the swallows have always come, he said. The loss felt personal, as if somehow a link had been severed between himself and the land.

'It is utterly devastating on a spiritual level,' he told me. 'I like to think I'm fairly in tune with what is going on with nature around us and the swallows are usually like something you can almost set your watch by every year. There are bigger swings of the pendulum these days, between hot, dry, cold and wet, and so many things seem to be conspiring against them.'

On the night of 5 April that year, a storm had whipped up over the Aegean Sea around Greece just as swallows and swifts were migrating north. Southerly winds pushed the flocks into air currents north of the Aegean that proved too powerful for many of the already exhausted birds to fly through. Over the following days the bodies of dead swallows and swifts were found in the streets of Athens, on seaside apartment balconies, Aegean Islands and a lake close to the seaport of Náuplia. The death toll, according to the local authorities, was in the thousands.

Were Chris Jones's swallows among them? Perhaps, but regardless his story tells us of something that rests at the heart of this book. How the great meteorological shifts of a rapidly changing climate are influencing the seasons as we understand and relate to them. How the weather, our weather, by which as an island nation we somehow define ourselves, is morphing into something beyond the reach of our cultural memory. And how the landscape and our own perceptions of ourselves are changing with it. The writer Richard Mabey defines our local understanding of the seasons as 'weather accents or dialects'; fine tunings between weather and habitat. And these accents are becoming blurred.

So much of any discussion of this falls prey to nostalgia, of course; a misguided sense that somehow the weather always used to be just so when that was never the case. But what is equally clear is that as the pendulum swings ever more wildly with each passing year of the twenty-first century, time can no longer be relied upon as it once could. Our connection to the seasons, and in the process a deeply rooted sense of self and place, is slowly being lost.

Coronavirus is a disease that has exploited our societal weaknesses. It thrives on the margins, exacerbating the political blind spots ignored by successive governments, picking off the old and the vulnerable, the frontline workers driving buses and making hospital beds, and proving particularly deadly for ethnic minorities. One of the lesser, and stranger, cruelties of the virus is the manner in which it deprives some sufferers of their sense of taste and smell, in some cases for months on end, I know of friends in the early wave of individuals to catch the virus who experienced those early weeks of spring as if watching through a window. They could feel the sun on their skin and hear the nascent birdsong but the essence of spring was somehow lost.

The aroma of the season is always its most powerful component, conjuring a medley of all that has come before. It is a sense interlinked with the amygdala and the hippocampus, the parts of the brain lodged deep in the limbic system that allow us to experience emotion and retrieve memory. The intoxicating scent of blossom in the air, fallen leaves mulching underfoot or rain after a prolonged period of drought all stir up these scattered pieces of our lives.

During that spring, unable to travel beyond the immediate confines of our homes and neighbourhoods, our minds raced backwards towards that muddled state pitched somewhere between regret and contentment where the British often find themselves most comfortable. We dug out old recipes and baked all the flour off the shelves. We gathered round the television to listen to an address by the Queen and spoke about conjuring the wartime spirit of the Blitz. We re-read old books and re-watched old television series. On one evening in June, the 1966 World Cup final was even replayed in its entirety. In the face of life as we knew it changing completely, we huddled under the comfort blanket of our own shared sense of nostalgia. This, aside from football, being our other national sport.

Lockdown engendered the curious paradox of the incarcerated, whereby the days drag and yet the months race by. Although in this respect I had something of a head-start. Life can enter into a similarly amorphous state when you are struggling to conceive – as at that point of lockdown my wife, Liz, and I had been for several years. It is a painful and occasionally crushing experience that alters your perception of time. The monthly cycle had become for us something to adhere to far more strictly than the seasonal one. Summers turned to winters and back once more and still we remain stalled.

Over these years attempting to create new life together I have found myself beginning to highlight aberrations in the seasons: an external way of giving meaning to the building sense of disquiet at something not quite right within.

CHAPTER TWO
Weather Watch

I am a weather watcher. For a decade I have written a weekly column on the very last page of the Saturday edition of the *Daily Telegraph*, documenting the weather of the British Isles. I was given the column not through any burgeoning talent for meteorology but because of my love of nature, folklore and history. 'Write about anything but the weather', was my editor's somewhat cryptic instruction when I first took over the page, particularly so bearing in mind the column was entitled 'Weather Watch'. 'And whatever you do, don't try and forecast anything.'

That, however, proved prescient advice. Whenever I have attempted to predict anything too precise, I tend to get it spectacularly wrong. My very first column, in fact, elicited a letter of complaint from a man living in Baltasound, the largest settlement on the Shetland Isle of Unst – the most northerly inhabited island in Britain. It was a warm weekend in May and I had foolishly written that largely unbroken spells of sunshine were forecast nationwide, bar 'those poor shivering Shetland souls lost in a persistent band of rainfall'. His missive was curt, pointing out contrary to my erroneous reporting the Shetland Isles had in fact enjoyed a balmy few days of marvellous weather. Like all the notes I have been sent by readers over the years, I still keep his objection in a folder. 'To readdress your Anglo–centricity,' he wrote, 'I suggest you visit the islands. Currently the wild flowers are at their best, the summer birds have returned, we will barely see any darkness until August. There could be no better place to be.'

Embarrassed as I was, I also could not help but feel a flicker of pride that my column had made it all the way to the Shetland Isles (albeit a day late: my complainant told me the Saturday *Telegraph* does not actually arrive on the island until Sunday, and he generally does not get round to reading it until the following Monday). Fortunately, it was not a mistake deemed serious enough to remove me from my berth.

'Anything but the weather', I have come to realise, means writing about how the changing seasons make us feel. It means tracing the literature, art and music our weather systems have inspired. It means trawling though folklore to describe the curious customs and rituals we have created in response to the weather. It means understanding the extent to which the weather we have grown up with has shaped each and every one of us. And in recent years it has meant noticing and reporting how far and how fast what we deem to be normal is changing before our eyes. That the seasons as we knew them are shifting, and perhaps disappearing for good.

My weather watching has taken me to the Faroe Islands to write about the solar eclipse, to Portugal to report on the communities destroyed in the wake of wildfire, and to the scene of the worst moorland blazes in living memory across the north of England during the heatwave of 2018, where I conducted interviews wearing a surgical mask handed out by police to prevent us breathing in the toxic air. On many other years I have trudged through the ruins of homes devastated in supposed once-in-a-century floods now occurring with alarming regularity.

There are far more qualified people than me to write about the science behind our changing climate and its influence upon increasingly extreme weather patterns. What interests me more is attempting to understand the impact of this unfamiliar weatherscape: how and why we choose to define ourselves by the seasons, even as our modern lives lead us to be increasingly disconnected from them.

Also, I intend this book to be a matter of record. Like all weather watchers I wish to document the minutiae of the seasons as I find them now. For as my local greengrocer (who no matter how many times I visit and whatever else is going on in the world will exclusively speak to me about the weather) describes the last light of a winter's day: 'Turn your back and it's gone.'

★ ★ ★

In T.H. White's *The Once and Future King*, the author described the Forest Sauvage where Merlin resided as a place with perfect British weather. In spring, carpets of wildflowers adorned the forest floor and in summer the sky was an impossible blue. In autumn the leaves turned crisp and curled and in winter just the right amount of snow settled. In short, White wrote, the weather in that mythical age did exactly as it was supposed to.

Of course, this was a sorcerer's vision, for it has never quite been so. We are as liable in this country to sentimentalise the seasons as we do so much else of our history, harking back to some glorious epoch and forgetting all the droughts, wash-outs and damp, dreary days along the way – not to mention the succession of brutal winters during the Little Ice Age which started in the fourteenth century and spanned hundreds of years, when the weather grew so cold frost fairs were held on the frozen River Thames. The fairy queen Titania summarised the unpredictability of the seasons in Shakespeare's *A Midsummer Night's Dream*. 'And through this distemperature we see/ the seasons alter: hoary-headed frosts/ fall in the fresh lap of the crimson rose'.

Living as we do on an archipelago surrounded by four oceans, battered to the west by Atlantic gales, to the north by the Arctic wind and to the east and south by whatever mainland Europe chooses to throw our way, the British weather will never be easy to predict. But what has become clear to me in writing weekly about its vagaries over recent years is that the seasons are changing far further than we might imagine and far faster than our assumptions of what should be doing what, and when, can keep pace.

It is not overly dramatic to envisage our cherished four seasons eventually reducing to two: a growing season and a dying season. Every decade in the modern era provides clear evidence that spring and autumn are lengthening and winter and summer are becoming increasingly muddied terms. Spring, for example, now arrives far sooner and moves across the country far more rapidly than it once did. Records show the march of spring from south to north occurred at 1.2mph between 1891 and 1947. Between 1998 and 2014 that increased to 1.8mph. Now, scientists say, it has sped up to 2mph with our flora and fauna racing to keep up. Autumn, meanwhile, is increasingly becoming delayed. A recent study by the Met Office estimated the 'growing season' to be, on average, a month longer during the past decade compared to between 1961 and 1990.

In 2020 the Australian summer was recorded as lasting seven weeks longer than in the 1960s and that same year in the south-east of the US spring arrived twenty days earlier. And while global temperatures are rising, freak weather events are becoming far more common. Since the dawn of the industrial age, the earth's temperature has increased by more than 1°C. The past decade has been the warmest on record and the next is set to surpass it.

We are the primary cause of this disturbing trend. After all, the fossil fuels we continue to burn and the carbon dioxide we emit into the atmosphere dramatically accelerates the rising temperature of the planet. Attribution science can increasingly establish clear links between escalating extreme weather events and human-induced climate change. 'Angry weather' is how the climate researcher Friederike Otto describes it in a recently published book of the same name. Otto, a physicist, philosopher and director of the Environmental Change Institute at Oxford University, is at the forefront of so-called attribution science. She and colleagues have developed precise weather models which map two worlds: one as we are living in today and one without all the billions of tonnes of carbon humans have pumped into the atmosphere. In contrasting the two they are able to pinpoint the extent to which humanity can be blamed for extreme weather events. It is a field of science that in time could lead to corporations and governments being held accountable in courts for their role in causing storms, heat waves and droughts. Otto and her colleagues have been described as the forensic scientists of climate change – a 'climate CSI'.

Are the seasons now a crime scene, unfolding evidence of our own wrongdoing? Certainly, our relationship with the weather is eroding into something more confrontational. I notice this in the increasing coarseness of the language we use to describe the weather. As we have grown disconnected from the seasons we have lost a vocabulary of wonderfully singular weather words that exist, or at least existed, in the

English language. This precise vernacular has been largely supplanted, nowadays, by reports of 'monster storms', 'scorching summers' and 'bone-chilling deep freezes'. Cold weather 'plagues' us. Rain washes us out. Newspaper websites are the worst offenders for such shock-jock reporting. Some publications' headlines even capitalise key words to emphasise the HELLISH weather on its way. Or think of another word that is rapidly losing all meaning: unseasonable.

★ ★ ★

In 2018, I volunteered for a year as a writer-in-residence for a charity on Sheffield's Manor estate. My work was about engaging young people living on the estate in the nature around them. Over the course of twelve months we used art and poetry to document the seasons as we found them.

One Saturday in October we conducted an experiment about nostalgia and the weather to examine how our previous experience might overpower that of the present. It was a muggy day. At the coastal weather station Donna Nook in the neighbouring county of Lincolnshire, 26.5°C was recorded, marking the warmest October temperature in seven years. I suggested we leave the room where we were writing weather poems and walk to the top of a hill overlooking the sprawling 1930s-built estate originally designed to move families living in one of the city's poorest districts out of slums and into social housing.

The weather shapes us and our communities in unexpected ways. Sheffield is one of the most unequal cities in Britain. In the south-west are leafy suburbs with Yorkshire-stone villas (once the residences of the steel magnates) and all the independent shops, cafes, galleries and good schools of any gentrified city suburb. Just a few miles away, meanwhile, are some of the most deprived areas of Britain (including the Manor estate). Health, education, opportunity all stand starkly divided. The underlying reason for this tale of two cities is supposedly down to the way the wind blows.

Typically, the prevailing wind across Britain lies somewhere between west and south. In Sheffield, this led the rich to build their homes where they did in order to avoid the soot and smoke from the steel mills blowing over the poor living in the north and east. I have read of the same theory applied to cities like London and Glasgow, whose east ends were where the labourers ended up living – and dying – cheek by jowl in the toxic waste of the industrial revolution.

The sky that October day was a mucky, murky grey. We sweated in our coats walking to the summit, where we were greeted by buzzing flies. There I asked the children what autumn meant to them.

They immediately started chattering back: slung spiderwebs shimmering in the low light; squashing a windblown apple under the heel of a wellington boot; a Mexican wave of trees turning to fire; raindrops pattering on the skylight at night; hot chocolate and the reassuring weight of a scarf coiled round your neck; dark nights looming like a shadow behind you; the smell of decay; the end of things.

The memories of the season implanted in the minds of those teenagers seemed incongruous compared to the humid heat. Here was autumn in name only, and the sentiments that word provoked sounded irrational compared to the reality before us. Isn't it strange, someone pointed out, that the silver birch and oaks had barely started to turn but their memory of previous October days was scuffing through great heaps of leaves in squeaky new school shoes? Which, we wondered, was the correct version?

Already that year we had worked through a stilted spring (the result of the swirling polar vortex that came to be known as the Beast from the East) and the joint hottest summer on record. By late July the ground had become so parched that the foundations of the old hunting lodge where Mary Queen of Scots was once imprisoned on the grounds of the estate (when it belonged to the Duke of Norfolk rather than Sheffield housing authorities) started poking through the yellow grass. And now we were left contemplating this temperate autumnal day.

What we had come to realise over the months writing and drawing the seasons was they were changing beyond our expectations and memories, and beyond what anyone might deem 'normal'. To the Celts the end of October marked the festival of Samhain, the finish of the harvest and the turning point into six long months of darkness. Similarly to those youngsters on the Manor estate it marked a threshold of sorts. But as we spoke about drinking hot chocolate and preparing to hunker down indoors, it was sunbathing weather on the Lincolnshire coast. Nostalgia, as the old saying goes, isn't what it used to be.

Sometimes readers of my weather column send me poems alongside their letters and I had saved one to read with the children that day: 'Song at the Beginning of Autumn' by Elizabeth Jennings, which describes the various sensations the season evokes. Jennings also considers how we give names to the seasons purely to create some semblance of outward form to our own moods. We formed a circle together at the top of the hill and read out a line each of the poem. The final line came round to me: 'When I said autumn, autumn broke.'

★ ★ ★

For as long as we have existed on this earth, humans have read the seasons through nature like runes. The word for this particular field of science was coined in 1853 by the Belgian botanist Charles Morren: phenology. The British pioneer of phenology was Robert Marsham. Described as that 'painful and accurate naturalist' by Gilbert White, with whom he enjoyed a lengthy correspondence, Marsham scrupulously recorded the seasons in his native Norfolk. He is best known for his volumes *Indications of Spring*, which he wrote from 1736 until his death in 1797, recording twenty-seven signs of the beginning of the season. He also kept careful notes of the winter of 1739/1740 – the coldest year on record – when temperatures dropped so low the contents of his chamber pot froze overnight.

A community of amateur phenologists continue Marsham's work. The Royal Meteorological Society coordinated a national recorder network which ran between 1875 and 1948, noting the emergence of flowers, birds and insects. Nowadays the Woodland Trust's Nature's Calendar project is the largest phenology network in Britain, having amassed 2.7 million public records spanning three decades. These are the little observations we all make every day, consciously or otherwise. From spotting the emergence of garden perennials to kicking a pile of leaves to listening to the lone melancholic song of a robin at dawn as you walk home following a night on the tiles, a sound that always sends my mind spooling back to my teenage years.

In recent years, the Nature's Calendar project has recorded red admiral butterflies spotted in Salisbury on 4 January, hazel flowering in Southampton in October (when it should occur in January), frogspawn, red-tailed bumblebees and budburst – all not happening when they should.

For Dr Kate Lewthwaite, citizen science manager at the Woodland Trust who oversees the Nature's Calendar project, it was the frogspawn that did it. She recalls being blown away by one particular sighting on 19 November when generally even the earliest appearance of frogspawn in the south-west is not until late January. Comparing the observations the Nature's Calendar volunteers have collected with the Met Office historical records has made clear the impact even a small change in temperature can have on the appearance of our flora and fauna. For every 1°C increase in any given year, she says, hawthorn will flower nine days earlier, the peacock butterfly appears five days earlier and the ladybird six days earlier. Over the past twenty years the dataset now shows 'spring' as a whole arriving 8.4 days earlier in Britain.

We are talking as we stroll through Londonthorpe Woods, which were planted close to the Woodland Trust headquarters in Grantham in 1993. It is the first week of November, a fortnight after that balmy day with the teenagers on the hill in Sheffield, and still autumn shows little sign of appearing. The birch leaves remain green, providing ample cover to the goldfinch darting between them, dogwood is still in flower and the sloe berries yet to fully ripen into their squashy purple globes. From deep within the trees we can hear (though not see) the rattling of several mistle thrush – known in folklore as storm cocks and the herald of the cold months – and yet a few moments earlier I had watched a drowsy bluebottle enlivened by the low sun dancing past a rowan tree.

'Something hits me nearly every year,' Kate tells me. 'Something where you just think nature has gone really wacky. I think the four seasons are too culturally ingrained to start tinkering with. But as conservationists we can start putting out messages to people saying, at this time of year it should be winter, but look all around you.' Behind her the canopy is a mass of verdant green: autumn stalled.

The Nature's Calendar project was started by a phenologist called Tim Sparks. From 1948 until 1998 there was no organised phenology recording until Tim, then a research biologist at the UK Centre for Ecology and Hydrology, started a pilot scheme that would be compatible with the historic records. He has been monitoring the changing seasons for the past twenty-five years. One of his most interesting studies focuses on a location of which I have first-hand experience.

In 2013, the same year I started writing my weekly weather column, I also wrote my first dispatch for the *Telegraph* from the Cenotaph on Remembrance Sunday – a ceremony I have reported on nearly every year since. What initially caught my eye among the massed bands of the Brigade of Guards and gathered members of the Royal Family clutching their poppy wreaths were the mature London planes lining Whitehall either side of Sir Edwin Lutyens' monument to the 'Glorious Dead'. 'The limp union flags billowed out from the rooftops,' I wrote in my piece for the following day's paper. 'The London plane trees that flank the Cenotaph shivered loose leaves on to the assembled crowds below.'

A few years previously the London planes at the Cenotaph also caught the eye of Tim Sparks, who noticed how green they seemed to be despite it being the first Sunday of November. Trawling through newspaper archives he collated as many photographs as he could of every Remembrance Day over the past century, from ceremonies in both London and Paris. The images depict a remarkable transformation. In 1919, when the first ceremony was held in London, the trees were almost totally without leaves. Later pictures depict similarly cold weather with Queen Mary watching from the royal balcony swaddled in furs.

But from the 1980s onwards, the trees have appeared increasingly green. In total, Tim has gathered forty-eight years of photographs spanning the century which, to ensure accuracy, he has presented to a panel to assess the extent of leaf cover on the plane trees. All the evidence, he says, points to delayed leaf fall in recent years. His study only runs until 2010, although, as I can stand testament in every year I have been present since, the trees often appear as if they have hardly shed any leaves at all. Tim contrasts the strange modern-day verdancy of the Whitehall trees with a popular Victorian poem written by Thomas Hood in 1844: 'No shade, no shine, no butterflies, no bees, no fruits, no flowers, no leaves, no birds! – November!'

You may wonder if it matters whether the trees remain in leaf beyond Remembrance Sunday. Or indeed that since the mid-nineteenth century Britain has warmed enough to be able to now accommodate a thriving population of ring-necked parakeets squawking down Whitehall during the two-minute silence. Life, after all, still goes on. But what fascinates me is that for as long as humans have existed on this earth we have sought to explain our lives by the passing seasons – the waymarks heralding the passage of each year. And now, within our own lifetimes, those very signs are falling beyond the reach of memory.

The Ancient Greek legend of Persephone marks one of the earliest human attempts to make sense of nature's oscillation between death and rebirth. The sweet daughter of Zeus and Demeter, the goddess of the harvest and fertility of the earth, was kidnapped by Hades, the king of the underworld, but every year she was permitted to visit her family for six months. That period marked the arrival of summer; her return to the underworld realm meant the onset of winter.

In the very simplest sense, centuries ago, when we farmed locally all that we ate, the weather was clearly a matter of life and death. But beyond that prosaic explanation, our connection with the weather ran far deeper. The proliferation of almanacs, symbols, stories and traditional sayings that persist within even our modern minds reveals the extent to which we have always lived by the seasons and sought to define ourselves through them.

Wild daffodil walks like Farndale in the North York Moors endure as one of the most vivid symbols of spring, yet the bulbs are appearing ever earlier. During the winter of 2018/19 a reader of my column sent me reports of daffodils flowering at the roadside on the A571 on Moss Bank, St Helens on Boxing Day. The daffodils, he admitted, could even have been in flower for several days prior to that.

If this pattern continues then by St David's Day, the feast day for the patron saint of Wales, the earliest blooms in parts of the country will already have begun to wilt and in some cases disappear altogether. What then? Do we move our days of celebration? Do we change our national symbols? Do we alter our lives, or carry on as before?

CHAPTER THREE

Storm Clouds

The noise wakes me first. A yawning roar rising in my ears. I wriggle down into my sleeping bag, trying to discern the sounds all around me: branches skittering against a drystone wall, the trunk of a large tree I had noticed the previous evening at the field's edge now groaning with the sheer effort of staying upright. Rain hisses over the outer layer of my tent and – snap – another guy rope is gone. In the half-light of a dawn not yet fully broken I can make out billowing shapes of my increasingly formless tent. The darkness crashes in, pawing at me from the outside.

Sleep is now out of the question so I continue to listen through the few millimetres of canvas that are all that stands between me and the weather. After a while the tinkle of loose tent pegs, the flapping of canvas and the water spuming through the porch give the curious impression

I am aboard a ship. I push my shoulders back into the rutted earth to assure myself I am still, in fact, in a pub car park in Derbyshire.

The storm stampedes all over the surrounding hills. At one point it clatters right over my tent and then – when the noise reaches a crescendo and I hunker down further into my sleeping bag in terrible anticipation of being lifted clean away – it gallops a few fields off. In his 1957 poem 'Wind' Ted Hughes described 'the booming hills' of a night-time storm, and during this brief respite when it seems to have shifted from above my tent the noise is more like distant artillery fire.

I think of a strange weather omen I had come across a few days previously while on holiday in East Neuk on the coast of Fife. I had been staying for a few days in the fishing village of Pittenweem where, in a municipal planter next to the harbour, somebody had placed the corpse of an eider duck. It remained there for the duration of my stay and each morning as I left the rental cottage I inspected its lifeless form for fresh clues of how it had met its demise, tracing its feathers and the green wash along its otherwise immaculate white neck. I wondered if it might have been a refugee from the nearby Isle of May, perhaps, which is a few miles offshore and home to a breeding colony of eiders. A duck that fled May to end up dead in a planter in early February.

The eider is nicknamed the Cuthbert's or Cuddy's duck after the legend of seventh-century St Cuthbert who, when the weather was foul, ensured the eiders had sanctuary on the Northumbrian isle of Inner Farne. The weather was curiously warm in Fife that week, but I wondered if the mystery death of the duck might be a portent of things to come?

A lone robin singing in defiance of the howling winds interrupts my thoughts; a wonderfully reassuring sound since I feel by now as if all life must have been torn asunder.

I pick my moment and crawl outside to meet the storm. I can see the blackened edge of the fire my friends and I had stood around the previous evening as the gales had started to pick up. Against its warm glow we had not even noticed the wind and rain pasting our jeans to our calves. I can see two of my friends have already packed up their tents during the night – or at least that is what I hope has happened – so there are just two of us left in the field.

I watch the grey light crack over a copse of trees along a disused railway track now turned into a footpath. Suddenly the sky turns a sickly mustard-gas yellow and the artillery begins to boom closer once more. In the rapidly thickening rain I grab the remains of my now jellyfish tent and sprint in the direction of the pub's outdoor toilet block. As I run, the tent opens like a parachute behind me: the wind hauling me back as if it has unfinished business.

I have been making this annual camping trip to the Peak District for ten years now. That year of the storm – 2020 – was the fiftieth anniversary of the 'February Camp', which was started in 1970. We are a ragtag bunch of attendees, generally a dozen of us, and mostly we do not even see each other aside from our annual pilgrimage on the first full weekend of February. It has become an immovable date in my calendar, part of the seasons we create for ourselves.

Depending on the year, the weekend on which February Camp falls sometimes coincides with the Celtic festival of Imbolc, held to celebrate the beginning of spring. Imbolc translates literally from the Old Irish as 'in the belly' and the festival was associated with Brigid, a goddess of fertility in Celtic mythology, later adopted by the Christian church as Saint Bridget. The Celts lit bonfires to herald the warmth slowly returning to the earth. We also unfailingly have a bonfire each year, primarily to grill vegetarian sausages and warm us up before we retreat to our tents.

I have gleaned the origins of February Camp from talking with the elders, some of whom still attend after fifty years, although they now sleep in lodgings at one of the pubs we go to rather than pitch their tents. The Camp was originally devised between two friends who had both become engaged to their respective partners in 1969 and wanted to secure an annual date in the calendar when they could still strike out into the countryside together. They chose the first weekend of February because they presumed the weather would be so bad that their new wives would not want to come. One of the founding members later died in a car crash, but the camping trip continued.

My own invite came via one of my best friends, Ben, whose father was an early attendee but died of cancer far too young, while we were still at university. They never made it to February Camp together, but these are traditions that get passed through the generations. Most years Ben can be relied upon to continue his father's legacy of temporarily losing his keys in the field where we camp.

After fifty unbroken years, February Camp holds great value as a scientific study of changes in the weather, though with our walks invariably taking us through numerous pubs this is no place for science. In between drinking and domino sessions we traverse the limestone ranges of the White Peak – 300 million years ago the bed of a tropical ocean (as another founding member used to unfailingly announce as he led us over the tops).

Reflecting on the ten years of camping out and listening to the stories that range far beyond my own experience, the focus often returns to the weather we have collectively endured. And also there is a quiet collective lament for what has passed so rapidly before our eyes. Despite there being two significant snow dumps during my own experience (in 2011 and 2019), generally the weather is unavoidably warmer and wetter. And during those drab years trudging under muddy skies through muddy fields, we walk back through our own weather memories with a powerful shared sense of nostalgia. We talk then of old bonfires lit with a single match on the crisp snow, sending flames crackling up into the starry night.

Storms have sometimes blown in, though few have proved as wild as that of 2020 – named Ciara under the Met Office classification. Later that day, sipping tea and eating cake at my niece's fourth birthday party in Manchester, my tent long packed away, I read that the storm we had slept through had grown in ferocity to become one of the most severe of the twenty-first century, wreaking a path of destruction across the country. A man in Hampshire died that Sunday after a tree fell on his car and more deaths would be reported over the ensuing days. Hundreds of thousands of homes had been left without power and swollen rivers had burst their banks. Gusts of up to 90mph were recorded; so strong that a transatlantic jet from New York's JFK airport arrived at Heathrow less than five hours after departure, breaking a subsonic record in the process.

My niece's party was interrupted by lightning and peals of thunder and we rushed to the back window to watch as the sky curdled and rain streamed down the patio doors. The internet router stopped working and the kitchen smart speaker was silenced. A beech tree was brought down at the end of the road and a team of council workmen arrived in a blaze of yellow flashing lights to chop up the trunk and clear the traffic.

I was three years old during the Great Storm of 1987. I still have two memories of the event: one true, one entirely false. The first, true, memory was walking around our local park, Highbury Fields in North London, in the weeks after the storm and watching a sculptor carve the huge trees that had been brought down in the winds.

The second, false, memory was that during the height of the storm while walking home from nursery I held on to a lamp post outside Highbury Magistrates' Court as the wind swept my legs away. I believed this for decades afterwards until I read a history of the storm and realised the winds were at their peak during the night when I was safe asleep in my bed. I mentioned my memory – so clear I can still see it now – to my parents who told me they had no recollection of their youngest nearly being blown away and liked to think that they would, if it had happened.

It was a moment to consider what other false memories we might unknowingly implant in our minds and how our expectations of the weather can affect our actual experience. Watching my niece, flushed with excitement and wearing a green polyester Tinkerbell party dress, I wondered how she might remember Storm Ciara and the toppled tree in the years to come, and how the weather she experiences might shape her own life.

* * *

Over the half century February Camp has taken place, the world's climate has transformed. Three dates in particular, each separated by several decades, serve to illustrate how. In 1968 scientists at the Stanford Research Institute delivered a report titled *Sources, Abundance and Fate of Gaseous Atmospheric Polluters* to the American Petroleum Institute. The report was one of the first to recognise the impact of burning fossil fuels on the atmosphere. It was intended to warn the trade body of the US fossil fuel industry of rising CO_2 levels which, if left unabated, could bring about what its authors described as 'climatic changes', including increases in temperature, melting ice caps and rising sea levels. The report fell largely upon deaf ears.

It would take another twenty years for the NASA scientist Dr James E. Hansen to provide what is considered to be the first mass warning to a wider audience about a term he had previously coined: 'global warming'. During a US congressional hearing on 23 June 1988, Hansen, dapper in a cream-coloured suit, declared 'with 99 per cent confidence' that a recent sharp rise in temperatures was not the result of natural variation but instead the product of human activity. The *New York Times* reported the following day that, assuming Dr Hansen's warning proved correct, 'humans, by burning of fossil fuels and other activities, have altered the global climate in a manner that will affect life on earth for centuries to come.' And yet again, we failed to act.

In 2018, fifty years after the Stanford scientists had delivered their initial warning, the Met Office published its first study of climate extremes in the UK as part of its annual climatic summary. Comparing the period between 2008 and 2017 with that of 1961 and 1990, the report confirmed that the forecast first delivered by the twentieth-century climate scientists had been proved broadly accurate: heatwaves in Britain are now lasting twice as long as they did fifty years ago, winter frosts are weakening and prolonged and severe rainfall is increasing.

The report found seven of the ten wettest years for the UK have occurred since 1998, with summers over the past decade 20 per cent wetter than between 1961 and 1990 and top temperatures also far hotter. From 1961 to 1990, the average longest warm spell each year was 5.3 days. From 2008 to 2017 this more than doubled to 13.2 days. The report also introduced a new metric, 'tropical nights', where the temperature never falls below 20°C. Previously such sweltering evenings were considered so rare there was no need to record them, but now they are an annual occurrence. One evening during the record-breaking summer of 2022 when Britain exceeded 40°C for the first time, Sheffield experienced its warmest night on record and the UK's highest minimum temperature ever of 20.5°C.

And all those dizzying figures supplying new pages of the record books is just the experience of the British Isles. Globally, the decade spanning 2010 to 2019 was the hottest on record, with temperatures now perilously close to the 1.5°C threshold above pre-industrial levels, which scientists have set as the precipice beyond which devastating impacts will occur (though the fear is now that we will actually far exceed this limit). As we enter the 2020s a NASA scientist has described the constant flow of records being broken as the 'drumbeat of the Anthropocene' – the present geological epoch named after the impact of the human race upon the natural world.

We can now forecast extreme weather events with far greater accuracy. The Great Storm of 1987, for example, was famously overlooked by weatherman Michael Fish. 'Earlier on today apparently a woman rang the BBC and said she had heard a hurricane was on the way.' So began the most infamous weather bulletin in British history. 'Well, I can assure people watching – don't worry, there isn't.' I interviewed Michael Fish in 2017 for an article marking the 30-year anniversary of the Great Storm. The then 73-year-old insisted to me it was an unfair stain on his reputation as the 115mph gales were not technically a hurricane. He conceded, however, that 'it did end up sounding a bit stupid.'

In total that night an estimated 15 million trees were uprooted across southern England. October 1987 had been a particularly wet month with the sodden ground already weakening the hold of tree roots. The fact that most trees had not yet shed their leaves for the season further increased their vulnerability.

To mark the thirtieth anniversary I decided to visit a stretch of woodland that had been decimated to see how life had recovered. I chose Ashenbank Wood in Kent, a stretch of 75 acres dominated by sweet chestnut trees. The oldest tree in the wood – a 350-year-old oak which overlooks the remains of a Bronze age barrow – was spared, but thirty years on the destruction that occurred around it remained all too evident. The trunks of huge chestnuts toppled during the storm lay mouldering on the forest floor. I paced along one fallen giant, being yaffled at by green woodpeckers to put me off my stride, and counted it as 60 feet in length – about 18 metres.

These trees will take centuries to decompose fully but new life had quickly taken root. Some of their branches had started to sprout vertically and form new trees out of themselves: life growing out of the ruins of the old. Similarly all the decaying wood has helped create a habitat for a nationally significant collection of beetles and some three hundred species of fungi.

The Met Office computer on which Fish and his colleagues relied in 1987 made four million calculations a second. Nowadays, the Met Office boasts a super computer capable of 14,000 trillion calculations a second. Most of the data comes from satellites that were rarely available thirty years ago. As a direct result of the Great Storm, deep ocean weather buoys were located around the British Isles to provide hourly weather information and monitor developing weather conditions. Where grid squares of 150km were once used for weather prediction, now the range is 10km.

By 2020 we were given plenty of warning that Storm Ciara would be barrelling in. As we look ahead we can see what is happening to our world with increasing clarity. And yet we remain seemingly powerless in the face of the changes that lie ahead. I wonder what world the February campers of another half-century will inherit. Will they look back on me as I do on the generation before and scream – 'Could you not see what was coming?'

★ ★ ★

A few years ago, during a visit to the Museum of English Rural Life in Reading, I came across a quartet of old Ladybird nature books about the weather, written by

twentieth-century biologist Elliot Lovegood Grant Watson
and illustrated by renowned wildlife artist Charles Tunnicliffe.
The books were published between 1960 and 1961, one for
each season, and entitled *What to Look For*.

They portray a world in perfect balance: weather, wildlife
and people all living harmoniously as the seasons progress.
In winter, geese feed on the surplus grain left over in the
stubble field while dormice curl up in their woven nests.
The countryfolk wear windcheaters and the cows sport
rough coats. Fieldfare flock on the cold north winds to feed
on the skeletal flower stems of cow parsnip and red campion.
Hoar frost ices this winter scene, illuminating the spider
webs hanging from the bare tree branches.

Sometimes when I pick up one of the books, I end up
reading all four in succession. I find them, and the idealised
world they portray, at once calming and yet laced with
melancholy. They convey a sense of post-war certainty that
we are masters of our own environment and the world will
always be so. Yet when I take this blueprint and spread it
over the reality of seasonal havoc we are now faced with
each year, I find these two worlds barely comparable.

Whole species have all but disappeared. The turtle doves,
which the summer book assures me I will meet on the
boughs of a horse chestnut in full blossom, have declined by
93 per cent since 1994. Where are the yellowhammer flocks
I'm told to look for in winter, and the snowy barn roofs that
look like wedges of Christmas cake?

The spring book contains a few snatches of verse from
the nineteenth-century poet Sydney Dobell's 'A Chanted
Calendar', which takes the reader through the seasons as
each flower comes into bloom. First comes the primrose,
then the windflower, then the daisies in May, then the cowslip
'like a dancer in the fair'. The cowslip was traditionally
scattered on church paths for late spring weddings or woven
into May Day garlands. A few years ago I sowed a few

cowslip seeds in a street tree pit outside my house and this February they have come into bloom already. By May their dried seed heads will already be fodder for the birds.

I can empathise with the poor cowslip's confusion. During the winter of 2020 it felt as if all the country was in flood and the weather changing so rapidly nothing could quite keep up. As I write this, yet another new environment secretary is in post following the general election and, like his predecessors, he is busy assuring voters that this time he really gets it: 'Climate, nature and people,' he writes in a Sunday newspaper article, 'we're all in this together.'

The crisply defined seasons of my Ladybird series, neatly quartered like an apple, are these days a mush. But sometimes I still come across a scene exactly as it is depicted in the books. Such encounters offer a tentative fingerhold on a world seemingly spinning out of control.

One example came the weekend before my Storm Ciara camping trip, at the Yorkshire Wildlife Trust reserve Potteric Carr, on the edge of Doncaster. Flanked by a busy A-road and overlooked by industrial parks, the reserve is often visited by starling flocks in winter, and that of 2019/2020 proved exceptional.

I visited with two old and dear friends: Daniel, whom I met in my first year at university, and his grandfather, Joe. After turning eighty, Joe started thinking about a 'bucket list' of things he wished to see and do, and observing a starling murmuration was chief among them. Over the years he has imparted a love of jazz upon me and so I was delighted to reciprocate by introducing him to this magnificent avian orchestra a few miles from my home.

Dusk was settling over the reedbeds when we arrived at Potteric Carr, whose name comes from the Old Norse, 'kjarr', meaning swamp or marshlands. It is one of the last surviving examples of how this South Yorkshire plateau would have looked before it was drained for agricultural land. For the starlings who still gather here, I suppose, it is a location that resonates in memory.

We stood outside as the sky reddened. For a while there was little sign of the birds, just a pair of swans gliding past the reedbeds and the distant roar of the traffic. Then, gradually, the starlings started to appear. My *Winter* Ladybird book records the birds arriving in 'converging flocks' towards their roosting place by the water, and so these Yorkshire starlings seemed to be streaming in from every corner of the county. As they met the main flock veered this way and that, like a snowball gaining in size and momentum.

Soon the murmuration was so vast that we could not fully take it in. It stretched hundreds of metres in the sky beyond our normal field of vision. The birds formed endless shapes: a breaching whale; a stretched limousine; a melting hour-glass; a map of America; a Wellington Bomber strafed with starling flak.

The *Winter* book claims the 'chatterings' of the birds can be heard from miles away but as they flashed over our heads they were silent, save the whirring of thousands of pairs of wings seemingly beating as one. I felt the rush of air during one such flypast, from which Daniel emerged with a jacket streaked with bird poo.

As the light faded and air grew colder, the murmuration drew closer together, crackling with intensity. Supposedly the birds gather this way in order to keep warm, as well as share information on suitable feeding sites. Suddenly they made the collective decision that whatever purpose they had massed together for was completed for the night. As if some unseen conductor had lowered their baton, the birds plummeted almost as one into the trees. Their performance over, night fell similarly fast. The first stars studded the darkness as we left the birds to their repose.

★ ★ ★

In 2018 a group of academics from King's College London and Southampton University published a joint paper considering the extent to which adverse weather conditions provoke feelings of nostalgia. The researchers claimed to be inspired by a line in Homer's *Odyssey* where Ulysses battled through Neptune's storm, propelled by a fierce desire for home: 'All earth took into sea with clouds, grim night/ Fell tumbling headlong from the scope of light'.

Participants were encouraged to listen to recordings of wind, thunder and rain and jot down how those weather events made them feel, as well as keeping a ten-day diary which was then compared against meteorological data gathered over the same period. The results were clear: adverse weather leads to psychological distress, which in turn leads to more intense feelings of nostalgia created by the mind as a soothing balm to transport us to a better place. 'Weather-induced self-regulation' was the term used in the research paper.

The seasons, after all, are a man-made construct, a collective yearning to erect a fragile scaffold around the year. The sixteenth-century poet Edmund Spenser delivered an early treatise on the idea of nostalgia and the weather in his 1595 poem, 'Colin Clouts Come Home Againe'. Written in Ireland following a return from the court of Queen Elizabeth I, Spenser's poem evokes a sense of pastoral longing of a landscape of wailing woods and silent birds. He wrote of how 'fields with faded flowers did seem to mourne' while 'the running waters wept for thy returne'.

The word nostalgia itself did not exist until almost a century later when Swiss medical student Johannes Hofer published his 1688 dissertation, *Dissertatio Medica de Nostalgia*. In his paper, Hofer describes encountering this curious disorder in particular among those whose occupations take them far from home, such as soldiers or servants. The symptoms he noted were similar to paranoia, except fuelled by longing, not perceived persecution, and similar to melancholy, except specific to an object or place.

The term is derived from the Greek *nostos* (homecoming) and *algos* (pain). Swiss soldiers were supposedly so susceptible to bouts of nostalgia (then viewed simply as a disease rather than a mental state) that the playing of a certain traditional milking song was punishable among the lower ranks.

Ideas of seasons, I have come to realise, are driven by nostalgia – an affliction to which I have always proved especially susceptible. They rely upon an idea rooted not in the present but in what has gone before and what might come again. Take Keats's idealised vision in September 1819 in his ode 'To Autumn' in which he wrote of harvest time in the lazy heat of late summer. By the next month thick snow had coated southern England, heralding the start of a long and notably brutal winter. By the end of December, ice floes disrupted shipping in the Thames Estuary.

As a child I adored this simple vision of the seasons, gleaning early satisfaction from the earthly visions it helped to conjure and seeking in it a sense of self and place. Long before I ever laid my hands on the Ladybird *What to Look For* quartet, among my favourite books was the *Brambly Hedge* series, also divided into four titles spanning the seasons, about a family of mice living in the countryside. Autumn bore an illustration of fat blackberries entwined about the cover; winter a roaring fire.

Other childhood favourites of mine were *The Hobbit* and *The Lord of the Rings*. When Bilbo Baggins departs in late September to embark on his adventure, the Shire had 'seldom seen so fair a summer or so rich an autumn', as if weather and landscape were conspiring to keep him from leaving all that he knew. Once into the borderlands an incessant rain strikes up, which elicits a powerful longing for home. So detailed are Tolkien's observations that when he received a screenplay for an animated film of *The Lord of the Rings* trilogy in 1957, he objected to the 'unseasonal depictions' of the weather in the script.

In 2013 a researcher at Bristol University actually created a complex meteorological model for the whole of Middle Earth. Mordor, the researcher concluded, would have possessed a climate most similar to Los Angeles, Western Texas or the more arid regions of Australia. The Shire, with an average temperature of 7°C and prone to bursts of gentle rain, was most akin to Lincolnshire – where my mum's side of the family is from and where the steady drizzle and salt winds evoke their own memories of childhood.

Having fought on the Somme in the First World War during hellish autumn downpours which reduced the trenches to a muddy morass and left the corpses floating in shell holes, Tolkien understood better than most the power of weather as a bridge to home. Leaving his wife to go and fight on the Western Front in 1916 was, he once wrote, like experiencing 'a death'.

While nostalgia serves its purpose in allowing us to relate to the weather and lament what has passed, the speed and scale of climate change requires a wider lexicon. The term 'solastalgia' was coined by an environmental philosopher called Dr Glenn Albrecht in 2003 to describe the detrimental impact on residents of an open-cast coal-mining boom in his native Australia. More recently it has been applied to victims of wildfires and floods. Albrecht defined the word as 'the pain experienced when there is recognition that the place where one resides and that one loves is under immediate assault.' He has boiled the term down further still, to 'a form of homesickness one gets when one is still at "home."' The writer Robert Macfarlane surmises the difference between the two terms. 'Where the pain of nostalgia can be mitigated by return, the pain of solastalgia tends to be irreversible.'

I experience this feeling in response to the weather: a fragmenting sense of place as the seasonal patterns I once presumed would remain ever so alter beyond recognition. But while I wrestle with a diminished sense of identity once entwined with the seasons, what of the natural world that has long existed in harmony with the weather? Memory provides a scant roadmap where anarchy reigns.

CHAPTER FOUR

Seasons Past

The weather frames our childhood memories. Autumn was kicking through drifts of plane tree leaves that accumulated in the north London park where I used to play when I was a boy. Summer was playing football until I was so hot and thirsty I would come home and drink what felt like gallons straight from the tap. There were snow days, too. One particular morning at primary school I remember being part of a dispirited gaggle of pupils whose parents still insisted we attend classes despite the dump outside. Spring, when I think of it now, always takes me back to my grandmother's beautifully neat garden stretching out the back of her Lincolnshire suburban semi. We would search for Easter eggs among the daffodils and at the very back was a wooden gate that led to an alleyway and, to my young mind, another world.

The other side of my family came from Yorkshire and I always felt a pull northwards. I spent the majority of my twenties living in various places around Yorkshire, starting my career as a reporter on local papers. My first job was as a trainee on the *Halifax Evening Courier* and for several happy years I rented a house with a childhood friend in the nearby milltown of Sowerby Bridge. We lived at the bottom of a precipitous hill on the edge of the moors in a triangle of stone-built nineteenth-century houses susceptible to damp and, according to some of our neighbours, other malevolent forces.

Our road was linked to the moors by Boggart Lane, named after the mischievous spirits who lived on 'the tops' and would scamper down in darkness to cause all manner of mayhem in the towns and villages nestled below. Occasionally curious creaks could be heard from the upper floor of our house. At certain moments when I was alone indoors I would suddenly, and inexplicably, feel the hairs on my nape begin to rise. There were other strange episodes: lit candles where both my housemate and I could swear neither of us was responsible, and one year our Christmas tree toppled over in the empty living room when we were both upstairs. One neighbour apparently could not even go so far as hanging a mirror in her living room without some evil presence dashing it to the floor. At least, that was how they told it in the local pub; I never dared ask her myself.

Most of the people our age in the town were heading in the direction of Leeds, Manchester or further afield, and could not understand our chosen rustic path. But life here followed different routines to the big city and I was captivated. Summer was a time for bilberry picking on nearby Norland Moor, whose rocky outcrops are carved with the initials of the workers who once escaped up here on Sundays for brief respite from the toil of the textile mills below. On warm days off I would carry a blanket and book to one of the wildflower meadows leading down from the moor. The seasons turned sharply in this part of the Pennines. Autumn drifted in on clouds of wood smoke indicating the long cold months ahead. It was not for nothing that we nicknamed the place 'Sowerby Fridge'.

My weather memories of that time are ingrained: walking among the ruins of the thirteenth-century St Thomas Becket church in the moortop village of Heptonstall in wind so fierce that it snatched my scarf from my neck and sent it slithering like an eel up into the sky. Or standing with some visiting friends at the top of Boggart Lane watching a storm roll in towards us down the Calder Valley, the clouds a coiled fury of electricity and pressure. After one particularly deep winter I remember arriving for a Saturday shift on a March morning and standing outside the entrance to the old nineteenth-century Courier Buildings (now sadly sold off), which was adorned with a magnificent clock, like the *Daily Planet* newspaper in *Superman*. I stood there feeling the sun on my face for what felt like the first time in months, watching it transform the dark sandstone around me into the colour of honey and warm some primitive feeling inside.

Before starting as a journalist and while at Leeds University I met the girl who would become my wife. Liz and I were friends long before we were ever romantically involved, both of us seeing other people, though I always hoped one day we might get together. We finally did, in the final days of the final term during a long hot summer in which I experienced a feeling that has never left: the thrilling contentment of finding the person you know you are supposed to be with.

We remained up north until the middle of our twenties but London, that human hoover, eventually sucked us in. I got a job as a writer on the *Daily Telegraph*, and Liz a position as a radiographer at Hammersmith Hospital. We lived in a flat in Finsbury Park overlooking a small patch of green space where a pair of song thrush dug worms in the winter months. We loved it there, but as the years passed our yearning to escape grew stronger.

On bad days I could shuffle out of our flat, walk down the roaring three lanes of traffic on Seven Sisters Road, spend thirty minutes underground to reach my newspaper offices above Victoria station, and then eight or more hours encased in steel and glass before embarking on the same journey home. Roughly, I once calculated, this routine permitted me about ten minutes outside in a single day. At least part of that time was accompanied by the station tannoys blasting out warnings about the 'inclement weather' I was never going to experience.

I remember one winter morning where I felt things starting to go particularly awry in my home city. I was walking down Oxford Street through a thick blanket of pollution. The sky was grey but it was still warm enough to be wearing a T-shirt among the throng. In the shopfronts the mannequins were wearing that season's winter coats, but the season itself felt like a fading memory.

In July 1665, the diarist Samuel Pepys walked the streets of London during the advent of the Great Plague. 'The season growing so sickly that it is much to be feared how a man can 'scape having a share in it,' he wrote in one of his diary entries that month. Pepys felt the sultry summer weather indicative of some great peril coming his way, and standing in that ghost of a winter I noticed a similar feeling taking hold.

Liz and I had always planned for our time in London to be temporary. After deciding to get married we agreed it was time to return back up north. The desire to own a house – something practically impossible for us in London – played a part in our decision to move to Sheffield, but its proximity to the Peak District was what truly appealed.

We both keenly felt the desire to return to a closer connection with the landscape and seasons that had been snuffed out by the seething metropolis. In a word, I suppose, to feel once more weathered.

★ ★ ★

Another desire propelled us back up north: a wish to start a family of our own. Like many of our generation we left it until we were in our early thirties to begin trying. We had moved into our new house in Sheffield and decided this was where we wished to stay put and raise a family of our own. But our bodies had other ideas.

Months slipped by and nothing happened. We assured ourselves there were plenty of possible explanations. I was busy, often working in London or abroad for what potentially were the key moments in the month to conceive; Liz had started a new and stressful job. We stayed relaxed and insistent that we would simply keep waiting for nature to take its course. And it did, as it always does, just not in the manner we hoped.

As the months turned into years we started to focus our efforts, using the various apps and tests which indicate the perfect time to conceive. We began to draw on unfamiliar calendars of internal temperature surges and blood. But no matter what we tried and how relaxed about waiting we forced ourselves to be, everything ended in the same creeping sense of disquiet.

January 2019 had already been a warm month. Down in London on a New Year's Day walk with my young niece and nephew we watched a pair of ring-necked parakeets nestbuilding in a hole in an old London plane in a park behind my parents' house and saw snowdrops poking through the earth. The winter blossom was out and the following weekend I noted premature budburst on the climbing hydrangea that swamps the back wall of my house in Sheffield. In previous years it had taken until the following month at the very earliest for the sepals to peel back and flowers begin to emerge.

I wrote about all this in my first weather column of the year and received numerous letters from readers who had spotted their own bizarre signs of the seasons gone awry.

Among them was a note from a reader who lived in Hampshire. On 30 December, she wrote, she had spotted a Cream Beauty crocus in her garden when normally the flower doesn't appear until March. The day after, she added, it was accidentally trodden back into the ground by a house guest. Also, in a south-facing sheltered corner of her garden, she had counted more than a hundred nasturtiums already well in leaf.

Later that January I was walking through central London on the way to a doctor's appointment on Wimpole Street when I noticed my first buff-tailed bumblebee of the year, drowsily buzzing about the base of a tree. Over the last fifteen years or so in southern urban locations, and in particular London, more of the bees have started to be spotted during winter. Previously queens would tend to go into hibernation until early spring but recent studies have shown the bees increasingly likely to wake up early, or just remain active throughout winter, due to the warmer weather and availability of nectar all year round. Some queens have even been reported establishing new colonies in autumn, allowing the worker bees to buzz on through the winter months.

The bee was a welcome distraction that day. I was walking to the doctor's surgery to submit myself for testing in the hope that the inner workings of my own body could be explained. I was given a plastic tube and shown my way to a starkly lit room with a bed whose blue mattress reminded me of a prison cell. The intricacies of my seed came back to me via a faxed pathology report divided into various sub groupings: motility, vitality, concentration and count. I opened and skim-read it in a manner which reminded me of collecting my school exam results, unfamiliar words flickering without much register in my mind as I raced to seek out any potential defect. My 'parameters were within the reference range' and 'motile sperm displayed rapid progression', all of which enabled me to breathe a sigh of relief and experience a surge of perverse male pride. But then I re-read the final criteria – morphology – which recorded an abnormally high number of defects.

The doctor said not to worry about the adverse finding too much. Supposedly the other high scores my sperm had notched up would counteract the dodgy morphology. He told me to try and avoid long periods of cycling, hot baths and drinking to excess but otherwise there wasn't much else to do. Keep trying and it will happen soon, he assured me. And eat more nuts.

In the weeks that followed the premature signs of spring around me seemed somehow heightened. These skewed seasons began to ring true for me in the knowledge that our bodies, too, were somehow not quite working as they should. And yet the thought of that winter bumblebee toiling on the stark grey pavement gave me renewed hope: that even in these preternatural times life can find a way of carrying on.

★ ★ ★

The early years of the eighteenth century coincided with a storm, possibly the most destructive one ever to strike the British Isles. Autumn in 1703 had blown in on a seemingly endless carousel of gales. Then, at around midnight on 26 November, an extratropical cyclone of abnormal ferocity smashed into the west coast of Britain from across the Atlantic, cutting a 300-mile path of ruin across southern and central England and Wales.

More than eight thousand lives were lost, including a fifth of the seamen of the sovereign fleet, drowned at sea. London and Bristol bore the brunt of the storm, with two thousand chimney stacks whipped off the roofs of the capital and Queen Anne forced to take shelter in the basement of St James's Palace. She later described the storm as 'a calamity so dreadful and astonishing, that the like hath not been seen or felt, in the memory of any person living in this kingdom.'

A few days later, the writer Daniel Defoe placed an advertisement in the *Daily Courant* and *London Gazette*, requesting first-hand observations of the carnage be sent to him care of his publishers. The first edition of his book, *The Storm*, was published the following summer. Defoe insisted on furnishing his work with the most 'authentick accounts we could from all parts of the nation'. In London, he noted, so many tiles were laid to waste that the price rocketed from '21s. per thousand to 6 [pounds]', while the great weather cock at Whitehall was blown down.

He included letters from correspondents across the country: a Gloucestershire vicar who described twenty-six sheets of lead blown off the middle of the church aisle, and a Cardiff landowner who counted hundreds of sheep and cattle perished. West Country orchards were flattened, whole woods torn up and barns blown down. According to the curate of the Hertfordshire village of Bishop's Hatfield, twenty large trees were levelled in the local park and a summerhouse by the bowling green razed to the ground.

The storm was the final instalment in a trilogy of disasters that had struck the capital during Defoe's lifetime: the third seemingly apocalyptic act following the Great Plague and Great Fire of London which occurred in 1665 and 1666, when he was still a child. All three episodes were believed by many at the time to have been incurred by the wrath of God for man's sins in what Defoe called the 'monster city'.

Ironically, the Great Storm marked the beginning of a gradually more settled period of British weather following the nadir of the Little Ice Age, which started in the Middle Ages and lasted until the nineteenth century representing the most pronounced climate anomaly of the past eight thousand years (save the modern climate crisis). Between 1570 and 1630, in particular, a dramatic cooling phase occurred as part of the Little Ice Age named the Grindelwald Fluctuation after the advance of the Swiss Grindelwald Glacier during that same period. During the seventeenth century in the northern and southern hemispheres, temperatures dropped by around 2°C. That is the same figure the modern nations of the world have pledged to restrict global temperature rises to during this century, through the Paris accord signed in 2015, and even within that narrow margin the impact of the Little Ice Age was tremendous, prompting near constant outbreaks of war, famine, natural disaster and disease.

During the coolest part of the Little Ice Age, some scientists believe the world entered into the new human-dominated geological epoch known as the Anthropocene. A recent study published in *Nature* focused on a pronounced dip in atmospheric carbon dioxide in 1610, captured in Antarctic ice records. This, researchers argued, occurred as a result of the arrival of Europeans in the Americas and a catastrophic collapse in the populations of Indigenous peoples due to the smallpox they introduced into the lands they invaded. An estimated fifty million Indigenous peoples died in a matter of decades and their farmland was reclaimed by the forest. The researchers named the 1610 dip in carbon dioxide the Orbis Spike, choosing the Latin word for 'world'. It was an event, they argued, that set the planet on the trajectory we find ourselves today and marked the moment the Old World collided with the new.

As the Little Ice Age progressed during the seventeenth century there was a marked increase in earthquakes, comets, volcanic eruptions and El Niño activity (the name given for when sea temperatures in the tropical eastern Pacific rise 0.5°C above the long-term average, impacting global weather patterns).

Some contemporary chroniclers sought to explain the apocalyptic weather through science. In 1671 Ralph Bohun, a fellow of New College Oxford, wrote a 'discourse concerning the origins and properties of wind' in which he described the atmosphere being altered by jets of water driven up from the sea in the manner of water whistling from a kettle. This was the Aristotelian view of the weather originally put forward in his treatise *Meteorologica*, in which winds flow about the earth through subterranean chambers.

Others believed the extreme weather was the product of God's wrath. John Milton's *Paradise Lost*, written during a particularly brutal winter and published in 1667, encapsulated this widespread sense of terror of the weather as summoned by a vengeful deity to punish us for our sins. One popular theory was that the earth was getting colder the further away the Garden of Eden was left behind.

In the middle of the Little Ice Age the Grand Duke of Tuscany organised an international collective, known as the Medici Network, which is regarded as Europe's first weather service. Running from 1654 to 1670, the network's aim was to record temperature in a dozen or so locations across Europe. The measurements were made using twenty-six newly invented Little Florentine Thermometers, a glass capillary tube sealed to a hollow sphere filled with a refined spirit and sealed with flame. Previously temperatures had been measured using the far less accurate thermoscope, invented by Galileo in the sixteenth century. When the first Little Florentine Thermometer was shipped over to England in 1661 it was received with rapture and gained the nickname 'the weatherglass'.

It was an invention that led to a renaissance of weather watching across Britain, one which coincided with the Little

Ice Age slowly relinquishing its grip over the course of the eighteenth century and the seasons gradually beginning to assume a more settled pattern. The most prominent of those to begin keeping lengthy temperature observations was the Oxford academic and medical researcher John Locke. In the late-seventeenth century Locke started his series of weather observations based on instrumental readings, which he submitted to the Royal Society. Outlining his hope for a nationwide network of weather watchers to provide similar standardised readings from every country, he wrote: 'Many things relating to the air, Winds, Health, Fruitfulness etc. might by a sagacious man be collected from them, and several rules and observations concerning the extent of winds and rains etc. be in time established, to the great advantage of mankind.'

Locke's vision was in fact rapidly taking shape as weather watching boomed among the literary classes. Magazines and periodicals soon started to include popular accounts of the weather. Locke, along with Luke Howard and Gilbert White, are among the more famous names, but local archives and family collections also hold a treasure trove of unpublished weather diaries written over the course of the eighteenth century.

It would take until the mid-nineteenth century before the Little Ice Age was deemed officially over, but long before that the increasingly settled seasons started to be widely regarded as a blessing bestowed upon Britain, one bound up with a burgeoning sense of national identity that emerged over the same period. The 1707 Acts of Union between England and Scotland had, after all, led to the creation of the United Kingdom in May of that year. Meanwhile the British Empire was expanding across the globe as the country raced to become the world's dominant colonial power. The English essayist Joseph Addison, co-founder of the *Spectator* (a daily publication in circulation between 1711 and 1714, not the weekly magazine of today), regularly touched on the English weather in his columns. In one such column Addison wrote: 'Nor is the least part of this our happiness, that whilst we enjoy the remotest products of the north and south, we are free from those extremities of weather which give them birth; that our eyes are refreshed with the green fields of Britain at the same time that our palates are feasted with fruits that rise between the tropicks.'

Against this backdrop, in 1730 the Scottish poet James Thomson published his famous quartet, *The Seasons*. It would go on to become one of the most commercially successful literary works in the English language. The first complete volume also contained Thomson's poem 'Britannia', a bombastic patriotic call to arms and a forerunner to his 'Rule, Britannia', later chosen as the national anthem of the new United Kingdom.

This new notion of the weather as an external projection of identity, somehow defining and elevating Great Britain as a nation state, tapped into a nascent public consciousness. *The Seasons* proved an instant bestseller and was continually published for the next 150 years. Practically every literate household possessed a copy, while children learned whole passages by heart. The book was extraordinarily influential on a whole generation of artists and writers. Wordsworth, Constable, Turner, Gainsborough and the other Romantics all drew heavily on Thomson for inspiration. John Clare was inspired to write his own poetry after being shown a copy by a friend. On a train to Linton in Cambridgeshire, Samuel Coleridge found a worn-out copy of *The Seasons* lying in a window seat. 'That,' he remarked to his travelling companion, 'is true fame!'

Thomson's great legacy was to articulate the way in which our weather forms our deepest sense of identity. The son of a Presbyterian minister, he evokes a landscape in each long narrative poem that is dependent upon God's grace. In 'Summer', for example, he sings of the sun 'in whom best

seen shines out thy Maker'. But he also understands the
pleasure of nostalgia in the weather. Perhaps his émigré
status as a Scot relocated to London helped Thomson
conjure the lost pastoral pleasures contained within his
poetry. 'In the wild depth of winter' he imagines 'a rural,
shelter'd, solitary scene where ruddy fire and beaming tapers
join to cheer the gloom.' In 'Autumn' he writes of the
pleasurable melancholy of the season: 'now the leaf, incessant
rustles from the mournful grove, oft startling such as studious
walk below, and slowly circles through the waving air.'

The 48-year-old James Thomson died at his Richmond
home of what was described as a low nervous malignant
fever. He had never married and left behind no family to
speak of, but in 1762 a statue was placed in Westminster
Abbey on the wall next to Shakespeare's memorial. Cast in
white marble it depicts Thomson in a loose robe holding a
book and the cap of liberty. At his feet are a mask and a lyre
and on the front of the pedestal is a relief of the seasons, to
which a winged boy points and presents a laurel wreath.

★ ★ ★

As well as exploring the links between weather and
identity — the sense that prevails today that the weather is
somehow *ours* — many eighteenth-century diarists were
similarly preoccupied with how the seasons could induce
actual physical effects upon us. The new eighteenth-century
fashion of documenting the weather led to a raft of accounts
analysing its potential impact on our health. Gilbert
White wrote in his letters of 'a miserable pauper' in the
village who had suffered from leprosy from an early age and
whose conditions appeared in rhythm with the seasons.
White described a 'scaly eruption' on the palms of his hands
and soles of his feet 'that usually broke out twice in the year,
at the spring and fall.'

Other weather diarists were practising doctors who became convinced of a clear link between climate and popular health. One such eighteenth-century physician, Dr Thomas Short, a tall, thin and hard-featured Scot who settled in Sheffield where he built up a considerable reputation for his work, maintained a weather record for four decades. Dr Short published numerous articles and letters on the links between health, air and the seasons, and believed his work to be at the cutting edge of understanding the human body. His work touched especially on the connection between foul weather and foul moods – something we now understand to be Seasonal Affective Disorder.

This association between atmospheric changes and the human constitution was in itself not new. During the Black Death of the fourteenth century, doctors at the University of Paris posited that the cause of the plague was an alignment between Saturn, Mars and Jupiter, which had resulted in 'lightnings, sparks, noxious vapours and fires throughout the air'. A miasmatic theory emerged from medieval medicine that diseases such as the plague and cholera were the product of bad vapours. Robert Burton's wildly successful *The Anatomy of Melancholy*, published in 1621 and reprinted five times in the author's lifetime, connected this medieval assumption of the dark vapours the air could wreathe within body and mind to the theory of humours, which assumed a healthy body exhibited equilibrium between four humours (blood, phlegm, yellow bile and black bile). Each was based on an element and each had its own season associated with it. Any external disruption, therefore, could alter a person's mental and physical state.

Many of the eighteenth-century weather diarists attributed exact physical and mental diseases to precisely what the weather was doing. One of the most detailed examples is Charles Bisset's *An Essay on the Medical Constitution of Great Britain*, written by the Skelton-in-Cleveland physician in

1761. Bisset divides the medical year into five periods, beginning with the summer solstice and finishing at the vernal equinox.

He claims Britain had managed to avoid some of the worst epidemics, which at the time were ravaging continental Europe, due to its position as an island nation 'fanned in its whole circumference by the pure and temperate sea winds.' This continual replenishment of fresh air promotes 'accretion, condenses solids and fluids and strengthens the whole body', rendering us more robust against disease than other countries. However, any unseasonal weather, he argues, such as close, wet summers or insufficiently cold winters, can give rise to all manner of bilious disease.

Bisset and his contemporaries drew clear links between atmospheric circulation and that of the body. During the period between the autumn equinox and winter solstice he deemed it vital to draw blood from his patients to protect them against what he describes as vapours, melancholy, madness, worms and rheumatic and acute putrid fevers. In his essay he describes one such annual practice: 'Towards the close of the autumn I have always drawn seven or eight ounces of blood from young stout men under a quartan … with remarkably good success: and in these cases the blood was often dense and sometimes a little fizy, or had some fizy spots on its surface.'

Frosts were welcome as they sealed in the noxious marsh gas – believed to be closely linked to all manner of ailments – and strengthened the vigour of the body in the same way we understand a cold winter to affect trees today by inducing a period of dormancy. During one frost-free and unseasonably warm spring in 1756, Bisset describes encountering all manner of dreadful diseases among his patients brought about by the insufficiently cool weather: slow fevers, shivering fits, hysteric affections and madness.

One young woman, he claims, had suffered the same symptoms each spring for the past five years and could only be cured of her fits by bathing in the sea. Another female patient was gripped by 'a moveable rheumatic humour, and with spasms of the muscles'. Mostly, he treated patients with a combination of what he described as moderate blooding: gentle carminative laxatives, anti-hysterics combined with temperature attenuates and diaphoretics. He also prescribed a variety of home-brewed treatments including a tea made out of rhubarb, senna, aniseed, Peruvian bark, roots of parsley and burdock, orange peel, saffron, and the leaves of mugwort and greater celandine, to be drunk two or three times every other day.

The eighteenth-century poet and weather diarist William Falconer elaborated on the idea of weather and mood in his 1781 *Remarks on the Influence of 'Climate' on the Disposition ... of Mankind*. Summoning the triumphant and blind patriotism of his age, Falconer embarks on a lengthy treatise as to how the weather affects our bodies and in turn our personalities, and why the British climate produces the very best of citizens. Supposedly, he argues, the cold British winters restrict the secretion of bile in the body and instead promote regular 'urinary and alvine evacuations', larger bodily bulk 'and humours less disposed to putrefaction'. In contrast to hot climates, which, he claims, engender cowardice, vindictiveness and timidity among those who live there, the character traits most illuminated by the British weather are steadiness of conduct, bravery and activity. Falconer admits, however, these can occasionally be diminished by violence, gambling and drunkenness, which increase with each degree of latitude travelling from the equator to the North Pole.

To the modern mind his observations seem downright racist and even some contemporaries disagreed with the extent to which the weather dictated our personalities and moods Writing his weekly 'Idler' column for the *Universal Chronicle* in 1758, Samuel Johnson argued; 'Our dispositions too frequently change with the colour of the sky. Surely there is nothing more reproachful to a being endowed with reason, than to resign its powers to the influence of the air, and live in dependence of the weather and the wind.'

* * *

And yet modern medicine has found the weather can have a dramatic effect on our bodies and minds. Several friends of mine suffer from Seasonal Affective Disorder (SAD), whereby their moods sink in the darker months. The link between seasonality and mental health was noticed by ancient scholars including Hippocrates and Posidinius, who wrote 1,500 years ago that 'melancholy occurs in autumn; whereas mania in summer.' The condition was not officially defined until 1984 by US psychiatrist Dr Norman Rosenthal who had studied a group of patients in the state of Maryland. Dr Rosenthal defined SAD as a recurring major depression with a seasonal pattern – with light, biological predisposition and stress the three most common causes.

People who suffer from SAD have difficulty regulating their levels of serotonin, which is responsible for balancing mood and melatonin, a hormone produced by the pineal gland, which responds to darkness. This combination of decreased serotonin and increased melatonin impacts circadian rhythms and leads to a rising prevalence of depression.

Most people register symptoms between the ages of eighteen and thirty, while women are said by Dr Rosenthal to be four times more likely to suffer from the condition than men. Those who live furthest from the equator in northern latitudes are most susceptible. A study conducted by Dr Rosenthal and a colleague found that in Florida, for example, the prevalence of SAD is just 1.5 per cent of the population, while in New Hampshire it is around 10 per cent. In Britain 20 per cent of us are said to suffer from winter blues and 2 per cent of us have been diagnosed with SAD. Long before it was officially recognised as a syndrome, William Blake created a monochrome print of a sullen figure perched on a cloud with his head buried in his hands. The 1797 illustration was captioned: 'On cloudy doubts and reasoning cares.'

I count myself fortunate in that I do not suffer the same affliction. Indeed I have always taken perverse delight in foul weather. But these days a similar though as yet ill-defined mental condition can seize me in its grip. It hits at various times: like during that sultry winter stroll down Oxford Street. Or the fact that as I write this the clocks are soon to go back and the trees in my garden still seem to think it is summer and have barely started to show the first tinges of autumnal colour.

I think back to the Ladybird Books of *What to Look For* in the weather and how skewed the modern reality has become compared to the post-war pastiche of harmony they conjure. Such unnerving signs of the seasons serve to constantly remind me of a disturbing truth: for all but one of the years I have been on this earth, there hasn't been a single month where global temperatures have fallen below average. The last such month was February 1985, when I was eight months old.

Over the years that we have been unsuccessfully trying for a baby, those weather observations have fed into an altogether more personal pain. What I have started to understand is that experiencing infertility is to discover the exact opposite of our relationship to the seasons. You come to live by a monthly sequence of hope followed by the inevitable anguish. Then another month passes and you steel yourself to face it all over again. These are the new calendars of the middle of our fourth decade. Rather than waiting for the earth to tilt upon its axis and the sun to cross the celestial equator as it does during the spring and autumn equinox each year, we are instead focused on more myopic cycles of egg release and fertilisation.

Various studies have attempted to explain the science behind possible links between the weather and fertility. Some researchers argue that the season you are born in can actually determine how fertile you are. One study from the University of Veterinary Medicine in Vienna claims that women with a birthday in July had 13 per cent fewer children than those born in December, while men born in autumn produce the fewest children and those born in springtime produce the most. The study concludes there is a possibility weather conditions (and the likelihood of catching infections) may affect foetal development at a critical stage impacting on future fertility.

Other papers have looked at the link between rising global temperatures and the decline in male fertility. Theories abound: mobile phones kept in our pockets; microplastics in the water we drink. One 2018 study published by the University of California, Los Angeles Institute of the Environment and Sustainability, claimed climate change, and the rising temperatures that come with it, is making it harder for couples to conceive. The study analysed eighty years of birth records in the US and found a significant decline in births nine months after any heatwave. This is not, the researchers argue, because couples have less sex in the heat (actually they claim it is quite the opposite) but due to the fact sperm production falls when temperatures rise. This, of course, remains a theory. But it reflects our continued fascination with how the vagaries of the climate can affect us.

My weather watching has helped in this regard. Studying the passage of the year, the rhythm of the seasons, has instilled in me a sense of being part of something beyond myself. 'It makes us feel the age of the earth, the greatness of Time, Space, and Nature,' wrote the poet Edward Thomas of his own weather watching. 'The fact that the earth does not belong to man but man to the earth.'

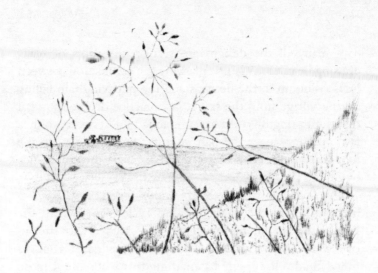

CHAPTER FIVE

The Changing Harvest

One summer weekend a few years ago, a friend was up visiting from London. The weather was as if painted by the landscape artist Eric Ravilious: chalk blue skies and splodges of double-cream clouds whisked up to cover the gritstone grey of the Peak District. On Saturday morning we decided to make the most of it, and pedalled our bikes up from the edge of Sheffield into the national park.

It was the season for well-dressing, a custom particular to this part of the world, where between May and August villages decorate their wells with ornate arrangements of flowers, bark, leaves and any other natural materials they can lay their hands on, set into a clay base.

The oldest example of the tradition is the village of Tissington, whose wells have been dressed by its inhabitants each Ascension-tide week for some 650 years. While these days many of the designs are secular, the roots of well-dressing lie in Christianity: marking the connection between Jesus rising up to the heavens and the purifying rain falling on the village from the sky. It is a practice deemed so vital that the Tissington wells have remained undressed on only a few occasions over the past century: two world wars, the foot and mouth crisis of 2001 and, in 2020, the emergence of Covid-19.

Our cycling trip was, if memory serves, in 2018 – a lifetime before the pandemic. We cycled on towards the Hope Valley, swooping through villages decked with streams of red and white bunting. Then, on the edge of a village called Bradwell, I spotted something that made me slam on the brakes. Here was Ravilious again. Or Constable. Or the Charles Tunnicliffe illustrations that filled my Ladybird books: a vision of pastoral perfection.

A family was hard at work in a field just off the main road. Four young boys of what I guessed to be between twenty-something and primary school age were each driving their own small tractors across a rolling field, strimming the dried grass to be rolled into hay bales. As they trundled along, clouds of insects drifted up around them and a squadron of swifts swirled overhead. Entranced, I watched the birds hawking low over the golden field, their shrieks (which led to them being known in folklore as the Devil's Bird) audible even over the rumbling engines. Despite being decked out in the ridiculous attire of the modern-day road cyclist, I could not help but intrude. I rested my bike against a drystone wall and self-consciously clattered over in shoe cleats and Lycra to introduce myself.

The mother of the boys came over. I said I hadn't seen such a gathering of swifts all summer and asked about the tractors: Old Massey Ferguson 36s and 65s, apparently; the boys' favourite toys. She was friendly and open but there was obviously work to be done so I quickly decided to get on my bike, as it were, and leave them in peace. Before doing so I explained I was a writer, and interested in such things, and asked if it would be possible to meet at a later date to hear a little about their lives. She gave me her name and number. I saved it on my phone as Sharon 'Hay'.

A few months later I drove over to the farmhouse where Sharon lives with her husband, John, and their boys. Their surname, I discovered, is Elliott; a name with deep roots in the area. I was welcomed into the dining room, all dark wood and cool stone, away from the glare of the sun outside, and offered a cup of tea. Sharon took a seat next to me but John, a stocky figure in a check shirt, preferred to stand at the head of the table with his arms resting on the chair. He steadfastly remained in that position for the next hour or so as he patiently answered my questions about their lives and work.

The Elliotts originate from the village of Hathersage, a few miles further down the valley. In the nineteenth century, John's great-great-grandfather, George Elliott, had one of four family farms split over two hills. They kept cattle then, as they do now. His grandparents ran the milk round in Hathersage, and John remembers his grandmother touring the village in a horse-drawn cart – later they upgraded to a van.

After their deaths the business was split between the family. John bought his farmhouse, which is surrounded by 10 acres. He also has other tracts of land he farms nearby and, further along the Hope Valley, four ancient meadows, designated as Sites of Special Scientific Interest due to the abundance of wildflowers that grow there. These are the vanishingly rare habitats which have been obliterated across the country by modern farming techniques, but that is not the way John and Sharon manage their livestock and land.

'We do farm slightly in the past, I don't deny it,' John told me. 'In simple terms it is all a slower process. Which doesn't suit all farms. There is nothing I like better than all the family in the field, involved and pulling their weight and having sandwiches when the sun is shining. That is how we've always done it.'

Meadow is a word derived from the Old English *maedw* from the verb *mawan*, to mow. It describes a process so rooted in humanity's psyche that it marked our first imprints on the land. The pasture is grazed by animals, but left in spring to grow and set seed. In mid to late summer the hay crop is cut by scythe, dried, then collected in bales. During the winter months, the livestock is brought back to graze the land and the process repeats. Over the centuries this has been the method through which we slowly shaped our countryside. By the eleventh century, notes the author John Lewis-Stempel in his book *Meadowland*, more entries were recorded in the Domesday Book for meadows than any other land use.

The Elliott family graze their land with a herd of around two hundred Limousin Cross cattle, which they rear for beef. The previous year they added two Angus bulls. John doesn't use fertiliser, nor does he spread slurry on the fields, in order to maintain the diversity of plant life. He avoids any overuse of medicine and proudly says that he hasn't been required to inject a calf to treat it for pneumonia for twenty years.

He grows the rich meadow grass for winter fodder and tells me they are one of the few remaining local farming families to bale it themselves rather than lease the work out to contractors. While many farms cut the hay twice or even three times each year, he will do so only once.

John has also maintained the centuries-old network of drystone walls across his land. The small Massey Ferguson tractors driven by his sons are all he will allow on the fields.

'My father used to say the bigger the tractor, the bigger the mess,' he told me. John is fond of the old sayings as well as the old ways.

Similarly, John deliberately keeps a smaller herd to avoid placing too much pressure on his land, even though it is a battle to do so and remain able to turn just a small profit. Modern consumer demand for cheap food has forced many farmers to utilise every inch of land. Farmers, John said, are facing 'terrific bills and are having to purge the living daylight out of the land to keep up with those bills.' He paused to look at the sunlight streaming through the narrow stone window frame on to the dining-room floor. 'It's as well we are farming in the past,' he said. 'Now if we had been stocked to what our land would normally take we would be in very serious trouble.'

He is talking about the weather. When we met, in early August 2018, the long dry spring had given way to a period of extreme drought. In Ladybower reservoir, not far from the Elliott farm, the water level has dropped so low that the drowned village of Derwent has emerged from the depths. The village was abandoned to the reservoir in 1943 and has only been spotted occasionally since, during periods of extreme drought. Before it was flooded much of the village was reduced to rubble, though the church steeple was left intact before later being dynamited. According to local legend, sometimes the church bell can still be heard tolling across the water. That summer, as the sun baked down and rain refused to arrive, it tolled for the livelihoods of many who worked the land.

John, who is in his fifties and has lived his whole life in the area, says the length and severity of the drought is unprecedented. Not a single drop of rain has fallen on his land since May. Sileage crops are down by a third, John says, threatening the supply of winter feed. A lot of neighbours have been forced to abandon their planned second cutting because nothing has grown since the first. 'I know people talk about the drought of 1976 but I have never known a time like it,' he told me. 'Suddenly ground that grows good grass is growing nothing.'

Talking to the likes of John and Sharon, who see these meteorological changes at a level of intimacy far beyond the grasp of many of us, I discern a growing distinction between weather and seasons. The latter is something they and generations of their family have built their lives around; a deep understanding of and respect for the calendar year and their place in it. From haymaking to walking his cattle off the high hills into the sheds on Lower Farm each winter for them to feed on summer sileage – a practice known as transhumance – it is work underpinned by a deep nostalgia and sense of how things should be. The strength of feeling he has for farming in harmony with the seasons is evident in his devotion to preserving what he has been taught and passing it on to his children. He defined it more simply as 'a love of the land'.

We stayed in occasional touch after that first meeting and I concoct a vague plan to visit John's hay meadows before they are cut the following year to see them in their full splendour. The next summer I was reading in my local cafe in Sheffield and came across the Edward Thomas poem 'Haymaking'. The Elliott family fields lifted off the page. The swift who circled the haymakers 'shrill shrieked in his fierce glee' across the sloping field and a sky of perfect blue. Thomas conveys a scene whose roots span millennia: from the nineteenth-century pastoral poet John Clare and the eighteenth-century pamphleteer and champion of rural England, William Cobbett, to the first people to farm this landscape some seven thousand years ago.

After reading the poem, I immediately phone Sharon from the cafe and she tells me that by chance they have arranged for the grass to be cut the next day. If I was quick, she and John would give me permission to go and walk on the fields that afternoon. I cycled home, picked up my binoculars and notebook and jumped in my car to drive the short distance from my house to their hay meadows: a mere 12 miles but a journey that transported me back through the ages.

The day before the cut was hot, and the Peak District still. I parked as instructed by John alongside a neighbour's cattle shed and walked into the first of his four fields over which a warm haze had settled. I picked up a fallen branch of an old ash tree bordering one of the fields and waded into the flowers.

As I picked my way through the waist-high meadow the names of the species I could recognise drifted through my brain likes lines of poetry: yellow rattle, white clover, purging flax, spires of sorrel flecked like crimson rust across the fields, purple pincushions of knapweed, common vetch, and creamy sprays of meadow sweet. The hedgerows at the field boundaries were comprised of old twisted hawthorn laden with vivid splodges of lichen, blackthorn, hazel, and snaked through with wild rose. Crickets churred and red soldier beetles clung to the flowering cow parsley while glossy black six-spot burnet moths droned past at knee height. I gave myself up to following a few of the moths as they barrelled through the meadow and as I blundered after them through the flowers became oblivious to where I was heading. When I emerged my bare legs were covered in scratches and my senses tingling, intoxicated by this meadowland.

I sat on the upper section of one of the fields looking out across the Hope Valley. A Manchester-bound passenger train trundled past in the distance and two buzzards mewed overhead. I felt a sense of sadness that tomorrow, all this life would be churned up under a tractor's blades but knew as well that the future health and biodiversity of the meadow depended on it. Life, death and regeneration are the underpinning of the seasons we have built our year around.

'Four seasons fill the measure of the year/There are four seasons in the mind of man' wrote John Keats in his poem 'The Human Seasons'. I found in John Elliott the embodiment of this connection, a man who has always known one way of life and fought to preserve it, not out of some misplaced sense of nostalgia but rather a love for the land. In spite of this dedication, even he recognises that relationship becoming strained like never before. He sees it now in every season. The rhythms of the year, ones he and his family have depended on for generations, are no longer what they were.

★ ★ ★

In 1949 a mosaic was excavated on Dyer Street in the centre of Cirencester, once an important Roman town in the south-west of England. Although no bombs had fallen directly on Cirencester during the Second World War, it was included in the country's peacetime reconstruction, with streets dug up and whole neighbourhoods rebuilt. The Seasons mosaic, as it came to be known, was identified as dating from the second century and decorated with busts of the four seasons: Flora, the goddess of flowers, wearing a garland with a swallow perched on her left shoulder; Ceres, goddess of agriculture, brandishing a sickle and a headdress of flowers and corn; Pomona, goddess of harvest, wearing a leopard-skin tunic and holding a pruning knife. Winter was missing, although is thought most likely to have been a hooded goddess holding a bare branch.

The Seasons mosaic is one of the earliest examples of our modern-day understanding of the four seasons: growth, harvest, decline and eventually death. In Ovid's *Metamorphoses* the four seasons are compared to man's four ages and to the four elements. While this philosophy would in time have a great influence on the way we perceive the seasons in the modern age, it has taken centuries to fully develop how we frame our weather in relation to ourselves.

Early Anglo-Saxon concepts of the seasons revolved instead around a dualism between day and night. The pagan Anglo-Saxon calendar was a lunar calendar which started on 25 December with Guili (Yule), the name for both the first and last month of the year. The extra days created by observing the waxing and waning of the moon were dealt with by the addition of an extra month. This followed closely in line with Norse traditions of understanding the world oscillating between two seasons: *sumarr* and *vetr* (summer and winter), and life and death.

For the Anglo-Saxons, October was known as *Winter-fyllep*, the beginning of winter, when a full moon marked the advent of six long months of darkness. It was common to count the years in winters, a person's age determined by how many they had survived.

Two seasons, *winter* and *somer* in Old English, cast the early year as a battle between light and dark. This conflict between the two appears in our earliest written texts. Winter is the season most referred to in *Beowulf*, while in 'The Wanderer' (the Old English poem preserved in the Exeter Book) the cold months frame the protagonist's lonely exile. This sense of dualism is also apparent in the twelfth-century poem 'The Owl and The Nightingale': 'You fly by night and not by day,' says the nightingale. 'I wonder about that and well I may.'

To modern minds the idea of the four seasons in Britain may seem sacrosanct but in fact spring as a defined period did not come into general use until the sixteenth century; autumn even later than that. According to Nils Erik Enkvist, a leading authority on medieval poetry, over time the Roman influence, combined with an observation of the

Equinoxes and the Christianisation of Western Europe, led to the gradual adoption of the four seasons throughout the middle ages. Enkvist suggests the French troubadour style had an increasing influence on medieval poetry in England throughout the thirteenth century, with poems such as 'The Cuckoo Song' reflecting a growing fashion for describing the passage of the seasons. Geoffrey Chaucer was the first to mention spring and autumn as distinct seasons, even if the unambiguous words for them did not yet exist. In Chaucer's translation of *Boece*, Enkvist counts six instances of 'the first somer sesoun' and two where 'autumpne' (borrowed from the French) is used but provided with an explanation so readers could understand this novel and ambiguous term.

Over the course of the thirteenth century, various Books of Hours – popular medieval devotional texts – also sought to define a seasonal calendar of sorts. January was a time for feasting, with several days of celebration observed throughout the Christian world. In cider-producing regions of England there was a variant of wassailing where people visited apple orchards to sing and toast the health of the trees by pouring drink on to the roots in order to frighten away evil spirits and ensure the prosperity of the year ahead. October was regarded as the time for sowing, and November for pollarding and feeding livestock on the last of the beechmast before slaughtering them in December. March was a time for pruning and turning the soil and the merry month of May for finding love. 'Come queen of months in company/ Wi all thy merry minstrelsy', wrote John Clare in his poem 'May'; 'The young girls whisper things of love'.

Since Roman times, Britain had followed the Julian Calendar (introduced in 45BC by Caesar), based on the 365-day year with an extra day every four years. Each year lasted twelve months, and 1 January marked the beginning of the year. To correct the lag of extra days this system inevitably entailed, in 1582 Pope Gregory introduced the

Gregorian calendar and ordained 5 October become 15 October. Typically, Protestant Britain proved unwilling to follow its European neighbours in adopting the new system and held out until 1752, by which time we had to correct our calendar by eleven days to bring it into line. This change was initiated by the 4th Earl of Chesterfield, who introduced a parliamentary bill to correct what he termed as the 'inconvenient and disgraceful errors of our present calendar.'

Beyond these official wrangles, writes the folklorist Steve Roud and author of *The English Year*, an annual calendar for ordinary men and women was emerging, formed by agriculture, religion and occupation. The clearest example of this is the proliferation of almanacs, which started from the mid-fifteenth century and soon entered mass production. By the seventeenth century, the volumes published were even outselling the Bible. Almanacs listed days of the week by date, month and fixed festival and combined details of astronomical events with the astrological forecast for the year. This calendar of proverbial lore, annual rituals and frequent festivities, argues Nick Groom in his book *The Seasons*, marked the progress of the seasons with absolute certainty.

Although folklorists question the extent to which pagan fertility rituals influenced our modern understanding and celebration of the English year, traditional rural customs followed this sense of rebirth and renewal. At harvest time a widespread practice across the British Isles was the weaving of small straw figures, which have latterly become known as corn dollies. Such figures went by a variety of names, including kern, kirn or churn, and were constructed out of perhaps the last stalks gathered in the harvest or the best straw cut. The figures, as children of the harvest, were hung up in the barn, in the hope that year's fortune would carry over into the next.

As society has urbanised over the course of the past few centuries, many of these traditional customs marking the calendar year have fallen by the wayside. Climate change and mechanised agriculture have blurred the lines by which we once read the seasons and rural communities have fragmented. 'Perhaps it is precisely because those shared reference points and memories are fading,' writes Nick Groom, 'that the seasons now seem so unstable.'

Yet even as we lament the demise of our cultural expectations of the weather, the changing variations also bring new opportunities. If we are living through the death of the seasonal cycle that has evolved over thousands of years, what, then, will emerge in its place?

★ ★ ★

The Derbyshire country estate Renishaw Hall stands a few miles south-east of Sheffield. Half a century ago from here you could see the distant smog billowing from the cooling towers of the steel mills while bitterly cold winters were a near annual occurrence. It was about as far from a traditional concept of wine country as it is possible to imagine, but that notwithstanding the then owner of the Grade I-listed hall, Sir Reresby Sitwell, decided to establish what was, at the time, the most northerly vineyard on earth.

Among other things, Sir Reresby was a wine merchant, and his family used to own a house with a small vineyard in the Chianti region outside Florence, which inspired him to set up his own back at Renishaw Hall. He spent considerable time touring different vineyards in Northern France, Germany and Switzerland to discover the hardiest vines that might cope with the freezing northern winters and often lacklustre summers. After conducting years of research, in 1972 he planted vines in the walled garden of the Derbyshire estate and on the lawns along the main drive leading up to his country house.

The wine, recalls Sir Reresby's daughter, Alexandra, who took over the estate following the death of her father in 2009, was famously awful: a sharp vinegary hit that would draw the gums back over the teeth. Local tenants winced whenever they were offered a glass as they came to pay the rent. 'When my husband used to come and stay, Father would crack open a bottle and I would stand behind him waving "Don't drink it,"' Alexandra told me over the phone one day. 'To say it wasn't drinkable was an understatement. It was more of a gimmick.'

When Sir Reresby first planted his vines, he was one of a handful of winemakers in England, all the rest of whom were established further south. But the rapidly warming temperatures of recent decades has turbo-charged the industry. In 2018, 15.6 million bottles were produced nationwide, more than doubling the previous record of 6.3 million set in 2014. The number of British vines has increased by 160 per cent in a decade to span 7,000 acres. In 2018, 1.6 million vines were planted with a further two million expected over the course of 2019. With Britain's top ten warmest years on record all having occurred since 2002, these days Renishaw Hall is not even the most northerly vineyard in England, let alone the world; that accolade now belongs to Norway's Lerkekåsa vineyard, a two-hour drive south-west of Oslo.

Whatever the Renishaw vineyard has lost in novelty value, it has gained in quality. According to winemaker Kieron Atkinson, the current tenant of the vineyard, climate change has created conditions in this part of Derbyshire similar to those seen in the Champagne region of France twenty years ago. He is now able to produce sparkling white, rosé and even red wines, which have won numerous accolades in the Decanter World wine awards and are on sale in local branches of Waitrose.

I arrange to meet Kieron for a tour of the vineyard one sunny August morning. He leads me through an orchard into the centuries-old walled garden of Renishaw, originally used as a paddock for racehorses before being planted up by Sir Reresby with a hectare of vines. Wood pigeons are perched on the walls greedily eyeing up the grapes slowly ripening in the sun. Roughly, Kieron estimates, the amount of fruit each vine can produce has increased from 2kg to 6kg in a decade. He knows the grapes are perfectly ready, he tells me, by a simple measure: the pigeons swoop.

One of the criteria for assessing the suitability of a wine-growing area is the so-called growing degree days (GDD), the measure of how often, and by what margin, the weather is warm enough for the vines to flourish. According to Kieron, at Renishaw they have progressed from around 670 GDDs in the 1970s to approaching 950 over the last decade – a rate roughly similar to Marlborough, New Zealand.

After the tour of the vines he leads me outside to another walled courtyard whose cobbles are bathed in the warm afternoon sun. The previous month, Derbyshire had recorded its highest ever temperature: 34.4°C at Coton-in-the-Elms, eclipsing the previous 34.1°C record set at the same location in 2006. That same week, 40 miles away on the other side of Derbyshire, residents were being evacuated from Whaley Bridge after the town's reservoir was damaged in heavy rainfall. I had spent a day reporting in Whaley Bridge, interviewing those outside the quarantine zone who were unsure when and even if they might be able to return. The year 2019 was proving to be one of those increasingly familiar summers when it felt as if the climate emergency was tightening its grip around us.

Renishaw that afternoon, however, felt a long way from such fearful times. That apparent distance is perhaps why as humans we have not been able to properly grasp the scale of the changes to the weather taking place around us. It is a golden rule of local news, which can similarly be applied to the weather: people are hard-wired to care far more about what is on their doorsteps, or immediately over their heads. When the sun shines, it is difficult to think about the unseasonal rain pelting down a few valleys away.

We opened chilled bottles of sparkling wine and something called pet nat, which is made in the traditional way by pressing the grapes with feet. 'It's difficult as a wine-maker to be profiting from what will be a global catastrophe,' Kieron admitted at one point. 'All sorts of stuff which is good for grape growing is bad for everyone else. From an insular grape-growing perspective, the fruit and quality from the wine-making process is far exceeding anything we could have done even a decade ago.' In forty years, he predicts, cabernet sauvignon grapes (typical of warm regions where red wine is produced) will be growing at Renishaw. And in southern England, I ask? 'They'll be growing raisins,' he says.

Aged in his early forties and a father of two, Kieron speaks in the matter-of-fact way familiar to many who have served in the armed forces. Before he studied viticulture at Plumpton College in Sussex, Kieron was in Afghanistan as a captain in the Light Dragoons. He arrived in the country in 2007 and was part of a unit conducting long-range reconnaissance patrols in Scimitar tanks. He credits his decision not to follow his fellow officers into the City after coming out of the army to his experiences in Afghanistan, and the awareness, accentuated by witnessing the loss of too many young lives, that our time on this earth is short.

The great joy of wine-making, he says, is the harvest. Each year he recruits more than sixty willing local volunteers to come and help collect Renishaw's grapes and briefly fill the old walled garden with chatter and life. His description reminds me of a documentary I watched one evening about hop-picking in Kent, one of those films made up of old newsreels with sparse commentary. During the twentieth century, whole families from the East End of London would travel out for their summer holidays to spend a few weeks collecting the harvest in the fresh air. Thousands made the journey each year; so many, in fact, that overnight trains known as 'hoppers' specials' were put on to transport pickers out of the capital.

There, amid the heady scent of an oast-house drying floor, was a particularly stark link between the seasons and fertility. Nine months after the hop harvest each year, there was supposedly a flush of unplanned babies born to hop pickers. From 1960, hop production declined precipitously in Britain. Machines replaced the pickers, and cheaper imports supplanted those grown in the old hop gardens of Kent. Of the 46,600 acres of hops that flourished in the county in the 1870s, only a few thousand survive today. In recent years there has been a small revival of the industry, with a few amateur hop-pickers once more heading out to help with the harvest, but they are a pale shadow of what has passed.

As Kieron talked about the harvests, he conjured a similar scene, though one not yet weighed down by nostalgia. It struck me as he described the changes taking place at Renishaw Hall that as warmer temperatures reshape our flora and fauna, they are similarly remoulding the English year. As the weather exponentially changes, it is inducing ripples through the land.

Harvest-time, Kieron told me before we departed, would be the first weekend of October. That was when the grapes were anticipated to be perfectly plump and the wood pigeons poised to dive down from their stone perches to feast upon them. And when I would be required to report for duty on a Saturday morning – secateurs in hand.

By the time I arrived back at Renishaw a few months later, the walled garden was already filled with volunteer pickers of all ages. Kieron, accompanied by his wife Jane and their two young children scampering about the vines, was handing out plastic crates and pointing out the best techniques of harvesting the grapes. I was set to work on a row of Seyval vines whose clusters of white grapes had already had their leaves stripped away to better expose the fruit. The sun once more was shining in Renishaw and as I worked I quickly got into a steady rhythm, snipping the bunches away at their stems and laying them carefully in my crate.

Spiderwebs laced between the vines sparkled with drops of early morning dew and I watched drowsy drunken wasps burrowing deep into the fruit. I spoke a little to my fellow pickers and most here had seemingly chosen the same reasons for their attendance as those that propelled the hop-pickers of the previous century: fresh air, a change in routine, a chance to be part of a magical word which eludes so many of us in modern life: harvest.

One family on the row next to me sang along to a local radio station on their phone as they gathered the fruit. At first I found my irritation towards the tinny music growing but then admonished myself. The point of joining this harvest, after all, was to experience a new seasonal activity, something that is not some insipid revival of a practice that modern life has left behind but rather one that demonstrates how our modern lives are adapting to the weather just as our ancestors once did.

And so against the drone of 'Angels' by Robbie Williams, I lost myself to cutting the grapes and filling the crates, the background noises fading away as the knees of my jeans became soaked through from kneeling on the dewy grass and my focus narrowed on the task in hand. Another snip and with sticky hands I laid one more squishy bunch down to be pressed into that year's wine.

We were finished by lunch, the crates of a successful harvest stacked up in piles all around us. Some stayed in the vineyard, unpacking sandwiches and sitting on the grass, just as John Elliott told me he loves to do with his family after a meadow has been successfully cut. I took my leave and went for a stroll in Renishaw's landscaped gardens.

On the back lawn in the shadow of the house I met the head gardener, a pony-tailed man in a green jumper who introduced himself as David Kesteven. He told me he had worked at Renishaw Hall for the past twenty-two years. We talked about the grape harvest and David recalled working with Sir Reresby to try and produce a drinkable wine. At one stage the wine was sent off to a laboratory to test the acidity levels and the readings came back as 'off the scale'. Chuckling, David said he can remember the comment that accompanied the laboratory results: 'Given time this may mature into a drinkable wine.'

When David took over as head gardener it was around the time when the 'hockey stick' graph had been published as part of an extensive study in the journal *Nature*, the first comprehensive attempt to reconstruct the average northern hemisphere temperature over the past thousand years. The study was based on numerous indicators of past temperatures, including tree rings, and showed temperatures remaining relatively stable until towards the end of the twentieth century, after which it suddenly shot upwards. The paper provoked years of statistical debate but its base conclusion remained unarguable: the rise in greenhouse gases due to human activity was causing this unprecedented warming of our planet.

David told me he did not need to cut into the tree rings of the old sentinels of the country estate to see how global warming was impacting on Renishaw Hall. He sees evidence of it all around him in every season – and not

just the award-winning wine now being produced The
year 2015 was, he says, particularly shocking. Crocuses
that would always flower in the top lawn around February
came into bloom before Christmas, and it has happened
every year since. His predecessor as Renishaw's head
gardener had always assured him that the best time for
daffodils was 26 April. Now, he says, they are in flower in
early March. The bluebell fortnight, at which Renishaw's
gardens would open to the public for two weeks straight,
has shifted forward from the middle of May to the first
week of April.

'Working here over the years I have just watched it
happen,' David told me. 'I've been in the same place, the
same environment – it's the perfect experiment and one I've
been watching for twenty years. Plants tell you what is
going on. If it's changing this much in a lovely place like
this, what is it doing where people are already living on
the edge?'

He mentioned he was so shocked by the changes he had
seen that he was going to stand as the Green Party candidate
for north-east Derbyshire. A few months later a general
election was called and by chance I was sent to report on
the count at nearby Bolsover, where David was standing.
It had become a key electoral battleground after local
polling had shown that the long-standing firebrand Labour
MP Dennis Skinner might lose his seat to a Tory, something previously deemed unthinkable in this old mining
constituency.

At 4am the assembled journalists were ushered into the
local leisure centre to hear the results of the count. The
87-year-old Skinner, nicknamed the Beast of Bolsover, who
had been MP here since 1970, had lost his seat to a 34-year-
old Conservative. Skinner was reportedly ill at the time
and had not come in person to hear the results, but I spotted
David Kesteven in the crowd, his Green Party rosette pinned

to his jacket. He had come fifth behind all the other political parties, including the Brexit party, with 758 votes. We nodded at each other grimly, and among the red and blue rosettes I thought of all he had told me about the landscape changing before him and how those warnings continue to be ignored.

CHAPTER SIX

Exodus

It is a hot, dry morning. Our footsteps kick up the parched forest paths and coat us in a fine dust that lodges in the back of our throats. We have been walking for hours, three of us strolling side by side through Thetford Forest clutching binoculars, cameras and a space-age-looking loudhailer to entice the subject of our interest. At dawn the forest had been alive with birds but as the rising sun has burnt away the last straggly tentacles of wispy cirrus clouds, the landscape has silenced and the bursts of our loudhailer are all that fills the forest. It crackles out intermittently, echoing through the canopy of the man-made plantation created a century ago. 'Cuckoo, cuckoo …'

This soon becomes a plaintive cry. No matter how many times we play it, the flesh-and-feather incarnation of that harbinger of early summer is nowhere to be seen, or heard.

The Brecks is a landscape sculpted by the weather. Belts of twisted Scots pine line the field boundaries like the ghosts of labourers bent double from their exertions. Also known as 'deal rows', the trees were planted in the nineteenth century as windbreaks to provide some shelter in this lowland country and prevent the shallow topsoil from being lifted clean away. In the past, the trees were carefully pruned into tight hedgerows, but following years of neglect they have burst free to tower over the flatlands, each wreathing its own unique pattern into the vast sky like acacia studding the African savannah.

Writers have long been drawn towards this otherworldly landscape. In Dickens's *David Copperfield*, the author described the area spanning some 400 square miles of dry heath and grassland between Norfolk and Suffolk as simply 'barren'. The Thetford naturalist W.G. Clarke coined the term 'Breckland', describing the constant need for farmers to 'break' new ground, cutting into the heather to grow their crops due to the lack of nutrients in the fallow earth. The seventeenth-century diarist John Evelyn described this shifting landscape as 'The Travelling Sands ... rolling from place to place, like the Sands in the Deserts of Libya.' It had been known, Evelyn pointed out, for entire Breckland estates to be consumed by this avalanche of dust.

The Brecks lays claim to being one of the warmest and driest parts of the country, but paradoxically is home to some of its hardest frosts. Even in the faltering cold snaps of the modern era, winter can still seal the landscape here solid as if some glittering plateau of concrete has been cast. And yet life still teems in this apparently hostile wilderness.

I once spent a day here spotting Breckland rabbits, more compact than our common species and boasting thicker pelts to insulate them against the bitter eastern winter mornings, darting through a spaghetti junction of warrens. The

Breckland rabbit was once so abundant that tens of thousands were slaughtered and dispatched to London each week for their meat and furs. Nowadays it is considered a species on the brink of survival, and one vital to helping preserve this rarefied landscape. In gnawing the grass and turning over the soil, the rabbits allow rare flora to flourish, such as perennial knawel, starry breck lichen, Spanish catchfly and spring sedge. These blooms provide nourishment for insects like the grey-carpet moth and five-banded tailed digger wasp, which, in turn, help sustain threatened populations of adder, common lizard, nightjar and stone curlew.

Over the past century, the Brecks has also been a stronghold for cuckoos. The 19,000 hectares of Thetford Forest, originally planted following the Great War to provide the nation with desperately needed new timber, has long provided ample nesting opportunities for cuckoos seeking to stow away their young. Since the early 1980s, cuckoo numbers have dropped nationally by around 65 per cent, but while they have vanished from much of the country the Brecks populations have been clinging on. Wandering through here on a late spring day one can still hope to hear that joyous and instantly familiar sound which inspired the famous thirteenth-century polyphony 'sumer is icumen in' and has long resonated through our culture. Except, it seems, that April morning in 2019, as I find myself tramping through the forest in vain pursuit of the cuckoo's song.

I have visited Thetford Forest with Paul Stancliffe of the British Trust for Ornithology (BTO) and a photographer friend of mine, Lorne Campbell. Thetford is the second stop for Lorne and me on a tour of the east coast for a series of newspaper articles documenting dwindling summer migrants. The previous day we had been in the North York Moors, the northernmost breeding ground for another visitor in catastrophic decline, the turtle dove.

Lorne and I had arrived at the North Yorkshire village of Lockton at sunrise that first morning to see a lone turtle dove who had arrived several weeks earlier and was hoping to attract a mate. The bird had been nicknamed 'Terry' by locals, who had quickly adopted him as a mascot of the village. Terry's, it transpired, was a tragic tale. Despite rousing himself each morning to flit between a gnarled Scots pine, ash and lime tree in the centre of the village, where he would display his black and white barred tail and coo a mellifluous mating call, there were no other turtle doves to answer. At one stage a lone female had arrived but she quickly paired off with another male, leaving Terry once more searching in vain.

By the time of our arrival, Terry had resorted, out of desperation, to attempting to woo a collared dove, an entirely separate species which unsurprisingly rejected his advances. There are only fifty breeding turtle dove pairs left in the North York Moors among two thousand nationwide (predominantly in the south and south-east). Later that summer Terry eventually returned unsuccessful to Africa.

The day after our turtle dove encounter we have a similarly early meeting with Paul at the BTO, whose headquarters is close to Thetford Forest. Lorne had driven down to Norfolk from the North York Moors that same evening and slept in his car for a few hours outside the BTO offices. I decided instead to return home to snatch some sleep before making the journey down from Yorkshire at about 3am. We meet up with Paul before sunrise and walk blinking and red-eyed into the Thetford dawn to discover that even here the cuckoo is making itself scarce.

★ ★ ★

It may only spend a few short months on the British Isles each year but no other bird is so closely associated with the seasons as the cuckoo. Its onomatopoeic refrain has for centuries been celebrated as the joyous sound of spring and early summer, when all life has reached its perfect peak and the very air seems laced with abundant promise. Two songs conclude Shakespeare's *Love's Labour's Lost*, symbolising the division of the year. Winter (or the Latin translation *Heims*, as it is described by the character Don Armado) is symbolised by the owl, and spring (*Ver*) by the cuckoo. The latter bird, Armado describes, appears on every tree 'when daisies pied and violets blue, and lady-smocks all silver-white, and cuckoo-buds of yellow hue, do paint the meadows with delight'.

As with the turtle dove, similarly celebrated by Shakespeare and which featured in Chaucer's *Parlement of Foules*, the cuckoo occupies a rich vein of folklore. Cuckoo fairs were once held across the country – and in places still are – to celebrate the arrival of the bird in April and early May. In the village of Downton, six miles south of Salisbury, the annual cuckoo fair is held on the first Saturday in May and a new cuckoo princess crowned by a cuckoo king before an archaic ritual known as 'opening the gate' ensues to allow the bird into the fair and ensure good weather over the following months.

The first Downton Cuckoo Fair was officially recorded in 1530 and continued until the Great War, when the custom faded. Unsurprisingly so, when one considers there are forty-four names that adorn the First World War memorial stone tablet on the north wall of St Laurence Church in the village, in memory of those long wartime winters when the men of the village never came home. The Downton Cuckoo Fair was eventually resurrected in 1979 and these days attracts tens of thousands of visitors.

So much value was historically placed on the cuckoo's presence that legends abound of people seeking to trap the birds in the belief it would guarantee fine weather all year.

Various so-called 'cuckoo pounds' are dotted about the country, some of which are named after monuments that date back to the Iron Age. In *Cuckoo Pounds and Singing Barrows*, a local history of Dorset, the author Jeremy Harte traces eight sites in the country named after cuckoo pens, or pounds, including at Corfe Castle and Langton Maltravers.

The pounds, supposedly, were constructed out of high walls and fences in order to better keep the cuckoo safely ensconced inside. The Wareham Spring Fair is one such example. Formerly held on 6 April, where the first cuckoo in Dorset was traditionally sighted, one year no bird was to be seen, an absence that presaged a foul and unseasonably cold spring. Legend tells it that the following year a fence was constructed around the first Wareham cuckoo that arrived. But the efforts to imprison it were futile and the bird inevitably flew away.

Stories of cuckoo pounds are early examples of man's recognition of the folly of attempting to control the seasons. The tales seem to originate from chapbooks: cheap and cheerful pamphlets that first emerged in the seventeenth century and were hawked at country fairs and on city streets. The Nottinghamshire village of Gotham is another supposed site of a cuckoo barrow. The location of Gotham is an interesting one, as its inhabitants once had a nationwide reputation for madness, even resulting in a book, *The Merie Tales of the Mad Men of Gotam*, published in 1565 and featuring a fenced-in cuckoo on the front cover. Supposedly this stems back to the reign of King John, when the monarch was planning to travel through Gotham and villagers feigned madness to persuade his courtiers to divert him past the village to prevent its road becoming a public highway. Word of Gotham's notoriety spread, so much so that the village inspired the name of the anarchic Gotham City in the comic *Batman*.

While failing to keep the weather permanently fine, there is some evidence that these cuckoo pens did serve some meteorological purpose. Harte points out that the entrenchment at Melbury Hill, Melbury Abbas, was used by people in the Shaftesbury area as a weather prognosticator. If the mound seemed to loom higher than usual, or was wreathed in mist, it was forecast heavy rain was on the way.

Searching around the British Library for weather stories one day I came across a mid-nineteenth-century pamphlet, printed by Cradock and Co. of Paternoster Row, celebrating the arrival of each migratory bird in Britain. The cuckoo, which according to the pamphlet appears in England on or around 17 April, features prominently. 'It is the hearing once again that dear voice, which seems the earnest of opening flowers and bright days, of flying clouds, warm gleans of sunshine, and all those indefinable delights which cling around our hearts in childhood,' the author wrote. 'Strange it seems that a single, monotonous, often repeated word should thus delight us. Yet so it is. The power of association is inexplicable.'

While the timing of the cuckoo's arrival was precise, its departure was shrouded in mystery. By the mid-Victorian era ornithologists had a growing certainty that the bird did in fact depart for warmer climes, but other theories remained under consideration – including the fanciful idea that cuckoos secluded themselves in tree hollows to see out the cold months until the following spring.

Back in Thetford Forest, Paul Stancliffe explains it is only in recent years that the mystery of cuckoo migration has started to be fully understood. The first vital clue came on 31 January, 1931 in a distant outpost of Cameroon when a cuckoo was shot by bow and arrow. The bird was destined for the pot, but its hunter noticed a small metal band around one foot bearing an identification number from where it had been ringed in Britain prior to embarking on its migration. Miraculously, considering the epic distances and communication issues involved, the ring ended up being mailed back to the BTO. Here was the first concrete evidence of the sub-Saharan migration of the cuckoo, but it would be another eighty years before scientists could fully trace its 5,000-mile return journey.

In 2011, with cuckoo populations plummeting, the BTO started a project satellite-tracking cuckoos in an attempt to explain what was driving their mortality. Of the five birds that were tagged, three headed south to the Congo Basin via Italy and two via Spain – the discovery of that second route was, at the time, new to science.

Since then some eighty birds have been satellite monitored, leading to the discovery that they arrive in Britain around late April and leave as early as the beginning of June – meaning they spend just 15 per cent of their lives in Britain. That finding somewhat contradicts the author of my Victorian migration pamphlet who erroneously claims the cuckoo, as diviner of the British weather, belongs primarily to us. 'England is her favourite resort, and her voice is heard here longer than elsewhere,' he wrote.

★ ★ ★

That year I had inadvertently followed the path of the cuckoo's migration. In March I had spent a few weeks journeying across the Sahel, the semi-arid belt spanning Africa south of the Sahara Desert, and whose inhabitants are the people on earth most disproportionately affected by the climate crisis. I had travelled there to traverse the route of the so-called Great Green Wall; an ambitious plan to plant trees right across this vast 5,000km region in order to reverse the rapid desertification which had led to 80 per cent of land being deemed degraded, making life unliveable for millions.

My journey started on the shores of Lake Chad, where the BTO-monitored cuckoos that travelled through Italy would stop off en route. I was there too late for the cuckoos who had already embarked towards the Mediterranean, but was still in time to see a magnificent hoopoe, with its slicked-back orange fringe reminding me of the climate-change-denying former US President Donald Trump, foraging for insects on a stretch of dried-up lake bed.

For years Lake Chad has been at the epicentre of a violent battle between the terrorist group Boko Haram and government forces. My first attempt to come here some years back had been cancelled due to a state of emergency being declared that meant even the UN (with whom we were travelling) could not guarantee our safety. The situation had calmed slightly since then but the terrorists continued to raid villages, take hostages, rape and pillage.

The people I spoke to here, many of whom were living in desperate conditions in refugee camps, driven out of their homes by Boko Haram and the periodic government counter assaults, spoke of the weather deteriorating alongside the security situation. In the Sahel temperatures have increased at nearly double the global average since 1970. That has resulted in prolonged periods of drought and the most ferocious heat I have ever encountered, nearing 50°C out in the desert. The day of my arrival I stepped outside without wearing a hat and within a matter of minutes began to feel the energy draining out of my body as if my head was in a microwave. In the face of such heat the very life of the land was draining away.

Over the past half-century Lake Chad – once one of the largest on the African continent – has disappeared, its waters receding by an estimated 95 per cent, leaving surrounding communities that once relied upon it for fish and water destitute. Here the seasons have always been divided into two: wet and dry, but as in the UK I discovered this historic pattern of life based around the weather was being violently disrupted. Whole villages were now praying to their own weather gods.

My Sahel journey finished in Senegal – the other route for cuckoo migration along the west coast of Africa towards the Mediterranean. The majority of the Great Green Wall had been planted here and there was optimism in areas where the trees had taken root: providing shade for community allotments and taking some of the fury out of the broiling heat. When planted in significant numbers, the trees have even helped create microclimates in the desert.

But amid these fragmented stories of hope, I also encountered whole villages where the young men had disappeared. Like the birds, they had left to travel to Europe in search of a new life. So many, in fact, that I heard a saying for those embarking on this perilous journey in Senegal: 'Get to Europe or die trying.'

On my very first day as a professional newspaper journalist in Halifax, my editor dispatched me on a dreaded assignment known in the industry as a 'death knock'. I was young, eager to impress, and so utterly unprepared that the editor, standing in the middle of the newsroom, took off his own tie and gave it to me to wear so I would appear suitably dressed when I turned up at the front door of the unsuspecting person in mourning. The story was about a helicopter crash which had claimed the life of a local businessman. His new widow opened the door to me and agreed to give me a few quotes 'in tribute' as we were taught to ask. I'm ashamed now to admit to the sense of exhilaration I felt as I raced back to the newsroom to receive the congratulations of my editor. Even more so when my story appeared on the front page the following day.

In the fifteen or so years since then I have reported on countless human tragedies: from the war in Afghanistan and the Ebola outbreak in Liberia to numerous refugee crises, terrorist outrages and murders. That initial burst of excitement I felt has long transformed into something very different and much harder to explain as I have soaked up what are often harrowing stories. I have always cried easily, but these days tears spring far more quickly into my eyes.

Few interviews, though, have moved me so deeply as the young woman in the northern Senegalese city of Touba whom I met a few days before I was due to return home. She was called Fali Ndiaye and was dressed in exuberantly patterned cloth, which belied her softly spoken words and sad eyes. We chatted on a balcony in the small compound where her entire family lived, and slept three to a room. Holding her wedding photograph in trembling hands she told me about her husband, Ibra, father of their two small children and one of 1.5 million people, predominantly men, to have left sub-Saharan Africa for Europe since 2010.

Ibra was thirty-two and struggling to make ends meet as a taxi driver when he decided to leave his wife of seven years and children behind to embark on a new life in Europe, where he might be able to send money back in order to support his family. He took a bus from Touba and travelled alone, hoping to end up in Italy.

It took six long months to travel up through Africa and scrabble together the money to pay people-traffickers, who agreed to ferry him across the Mediterranean. Occasionally Ibra would phone home. The last time was a week before his death and Fali said her husband asked for her forgiveness for failing to send any money back since his departure.

Ibra tried to remain positive during that conversation, saying that he was soon to sail across to Italy, but Fali admitted that they both had a terrible, if unspoken, sense of an ending. She discovered her husband's fate on a hot and dusty Monday afternoon. A friend he had met on his travels telephoned Fali from Italy to say the dinghy they had been travelling on had capsized when they were in sight of land. Those on the starboard side had managed to swim free and make it to shore. Everyone sitting on the port side of the dinghy, including Ibra, had drowned. 'I will never forget the day of that phone call,' she told me. 'I feel so lonely without him.'

The same climate crisis that is rendering swathes of the earth increasingly inhospitable and driving so many young people to their deaths, is causing a similar decline in migratory bird populations. For cuckoos, conditions have worsened all the way along their migration routes between Africa and Great Britain, from habitat loss to the decline of the insects upon which they feed. Along with another migratory bird in rapid decline, the swift, cuckoos are yet to speed up their arrival time to coincide with the earlier onset of spring in the British Isles.

It has been suggested that one factor behind the decline in cuckoo populations might be the shifts in breeding patterns of the host species whose nests the cuckoos invade to lay their own eggs. The main British hosts are the dunnock, meadow pipit, pied wagtail and reed warbler, which have shifted their breeding forward by roughly a week since 1994 due to warmer winters and earlier springs. However, scientists have analysed this early breeding for its potential impact on cuckoo populations and concluded it does not make a tangible difference. A more likely reason is the lack of prey available during the breeding season due to insects failing to adapt to changes in the weather and the increasingly arduous conditions the birds must navigate to arrive here.

Back in Thetford Forest the lack of sleep has caught up with us. After watching an exciting rarity for me in the form of a chirruping yellowhammer perched on the trunk of a dead tree, I lie on my back on the grass in a forest clearing staring up at the bright blue sky and listen to Paul talk about cuckoos.

We are searching for a particular cuckoo, named PJ, which has been tagged since 2016 and by that year had been followed over three successful migration loops. In 2019 he had arrived back at Thetford on 26 April from his wintering grounds in Angola. After studying PJ for so long, Paul admitted he had developed a close personal attachment, tinged with concern that the birds are becoming far harder to spot.

Such cautionary delight at the cuckoo's arrival was echoed by Chaucer in *The Canterbury Tales*. 'For when that they may hear the birdes sing/ And see the flowers and the leaves spring/ That bringeth into hearte's remembrance/ A manner ease, medled with grievance, mingled with sorrow/ And lusty thoughtes full of great longing.' We have always invested much hope in cuckoos. For a bird which bears sunshine on its wings, there remains the fear that one year it may simply not return.

After resting for a while we decide to return into the forest for one last attempt. Strolling through a plantation of conifers we emerge into another clearing studded with the trunks of dead trees – perfect singing posts for cuckoos. Then we hear it, the instantly familiar metronomic mating call of a male cuckoo, reverberating about the forest. It takes a while to spot him, singing from the thick boughs of a conifer, and as soon as we do so he is off. As he flashes past his barred underparts reminded me of those old comedy burglar outfits, which is rather fitting when one considers the surreptitious breeding habits of the cuckoo, who pilfer nests to their own ends.

Walking on and a few minutes later Paul identifies the raucous babbling of a female cuckoo, a very different sound I would not otherwise immediately have recognised. Part jay, part bathwater gurgling down a plug hole (as Sir David Attenborough once described). The pastoralist poet John Clare preferred a less prosaic description of the female cuckoo: 'Still waking summer's melodies/ and singing as it flew.'

★ ★ ★

Migratory birds plan their journeys along something called 'seasonal pulses'. They want to raise their young at a time when flowers and insects are flourishing and food is abundant. In late 2019 the BTO published a report assessing the long-term impact of climate change on bird populations. Of the sixty-eight species analysed, one-third were proven to have been affected by the weather. In five of the eleven English regions studied, cuckoos were the bird that registered the biggest decline. However, north of the border, Scottish cuckoo populations have remained relatively stable since the 1990s. Indeed in many places cuckoos are actually increasing. It is here up north where the cuckoos find the breeding conditions most like those they are used to.

In Britain a north–south divide exists even when it comes to the weather. Since the 1970s Britain overall has warmed by roughly 1°C but meteorologists have found that temperatures in the south of the country have been rising more rapidly than those in the north. One widely publicised paper on this inconsistent impact of climate change found that since 1950, the hottest days of the year have risen by at least 2.5°C in much of the south-east, but just 1°C or so in the north-east.

As the weather changes, all life has started shifting northwards. Insects in the warmer southern half of Britain have been moving up the country in search of pastures new. Trying to somehow envisage this is an exercise in formication. Beneath our feet and winging about our heads a largely unseen exodus of climate refugees is on the march north at a rate of around 5m every day. Like any displaced species escaping drought and deluge, their motives remain simple: food, security, a place where the weather and the nature it helps dictate might somehow make more sense.

A recent study published by a group of scientists at York University analysed 25 million recorded sightings of some three hundred invertebrate species gathered over forty years to better understand this northern advance, and discovered some species, such as the emperor and migrant hawker dragonflies, have sped northwards at between 17 and 28m per day. Another species rapidly journeying north is the comma butterfly, which has spread from Yorkshire to Aberdeen.

The research was led by Professor Chris Thomas, director of the Leverhulme Centre for Anthropocene Biodiversity at York University, which works to establish the winners and losers in this new geological epoch. He has been studying this largely hidden migration for decades and in 1999 co-authored a paper published in *Nature* which was the first large-scale evidence of poleward shifts in entire species. Of the thirty-five non-migratory European butterflies examined, 63 per cent had shifted northwards by 35 to 240km throughout the twentieth century, and only 3 per cent had shifted south.

In Britain, the speckled wood butterfly was one such species to have shown a dramatic expansion northwards. In order to determine the extent to which the warming weather was driving this shift, Professor Thomas and his team selected species that were less likely to have been driven away by modern alterations to habitat.

In a separate paper co-authored by Professor Thomas on a similar theme, the brown argus butterfly was found to have spread northwards by 79km in the UK in twenty years, a figure 2.3 times faster than the average expansion rate for species globally (which is around 16.9km per decade). The butterfly was thought to have reached its northern range boundary in Britain in the 1980s, largely restricted to using rockrose as its larval host plant in calcareous grasslands. Yet since then the brown argus has rapidly extended its distribution northwards, colonising large areas where even rockrose is absent.

It is unclear how much the northern spread of the cuckoo is to do with climate and how much habitat loss, says Professor Thomas when I arrange to speak with him over the phone one day. After all, it may be warming up in England but the bird's global geographical range lies far further south.

Still, there are numerous other examples of established patterns changing with the weather. Overwintering blackcaps are now rising exponentially – having increased fourfold since the 1970s. Ring-necked parakeets have marauded up the country from London to inner-city Glasgow where they have established a colony believed at one time to be the most northerly flock of parrots in the world.

There are, nowadays, regular reports of swallows staying here over the winter while the likes of goldfinch and goldcrest, typically susceptible to winter frosts, are thriving at garden feeders. In December 2019 I recorded in my diary the first ever sighting of a goldcrest in my Sheffield garden, hopping among the upper boughs of an acer the woman who previously lived here planted for her son. As I watched Britain's smallest bird flitting between the helicopter seed-cases dangling from the tree I was reminded of an old fable I had read about goldcrests hitching a ride on the backs of larger migratory birds that arrived here in winter from northern Europe, hence the old nickname, 'woodcock pilot'.

★ ★ ★

Migration stayed on my mind long after that springtime hunt for the cuckoo. The first blooms wilted into the long lull of late summer and I still found myself dwelling on the idea of seasonal pulses. Having watched the spring arrival of birds on to these shores I wanted to see the reverse. And so one late September morning, as the blue-veined night marbled into a red dawn, I hurtled down the eastbound carriageway of the M62 following the invisible flight lines of the autumn migrants in the sky above.

The shipping forecast was on the radio, the familiar metronomic beat of Viking, Forties, Tyne, Dogger ... and then the name of my destination, Humber. I drove past the suspension bridge and the glittering docks of Hull and on to a spit of land that curves out into the estuary like a scimitar blade. This strip of saltmarsh, surrounded by mudflats and water, is known as Spurn Point. It is the first stop for birds arriving into Britain over winter, and for those departing back across the North Sea, the very last sight of land.

It was still dark when I arrived at the Spurn bird observatory, but two young volunteers were already picking their way through last night's moth trap, examining the specimens they had captured in a device fashioned out of old egg cartons placed inside a wooden box. We had only just started chatting when my guide for the morning, Rob Adams, chairman of the observatory and a fixture of Spurn, appeared grinning as he pedalled an electric mountain bike from the direction of the caravan he had been sleeping in.

After an enthusiastic handshake and some unsolicited news bulletins – five thousand meadow pipits sighted the previous day – we scaled a watchtower to gain a view over the flatlands. Here, surveying all around us, I quickly discovered a conversation with Rob in migration season is a distracted affair. His history of the observatory (which

opened to visitors at Whitsuntide in 1946) was interspersed
by a continual roster of sightings: roe deer, a skein of geese
flying far above us, a pair of meadow pipits chirruping high
in the sky. At first I found it an effort to keep up but soon
felt myself being oddly soothed by the diversions of his
linguistic wanderings.

Nothing stays in Spurn for long, Rob said. Well, the birds
at least. He has been here on and off since 1969. Even
migrants that arrive after a battering 20-hour flight across
the North Sea will touch down on the hawthorn thickets
for only a few minutes before taking flight again. The
migratory urge is very strong, he said, but news of a captured
treecreeper in the nets crackling over his walkie-talkie
interrupted his train of thought and we hurried back down
the ladder and over to a green shipping container grandly
termed 'the ringing laboratory'.

Considering we were reaching the peak of the autumn
migration it had been a relatively quiet morning. Heavy
rains the previous night and the wind blowing in the wrong
direction meant the vast flocks that had been sighted in
recent days had subsided. The weather here is everything.
When the wind is blowing from the east and cloud cover
thick and low, tens of thousands of birds at a time can
suddenly tumble out of the sky like a snowstorm.

According to Rob, things are now changing with the
weather, and old-established timings can no longer be relied
upon. Spring arrivals are coming much earlier than usual
and autumn migrants staying around longer before they
depart for warmer climes. A swift, the symbol of Spurn, had
been sighted that same morning. Typically, the birds should
have already gone south by mid-July.

Various mist nets were slung about the observatory site
in order to catch birds so they can be ringed. It was a
set-up I had seen once before, albeit in very different
circumstances. Some years previously I had visited Cyprus

during the autumn migration in order to report on the
annual slaughter of hundreds of thousands of songbirds as
they pass over the island. The birds are killed for an illegal
but widely eaten island delicacy called ambelopoulia,
where they are served a dozen at a time, boiled, pickled or
fried whole.

To catch such industrial numbers of birds, the trappers
would erect mist nets at night and play birdsong rigged up
through giant speakers to entice the birds, which would
either become entangled in the nets or get stuck to branches
that had been glued.

I spent a few days with a group of activists working at
night to cut the nets down and free the birds. We met at
midnight in a car park on the outskirts of the party town of
Ayia Napa and crawled through farmer's fields patrolled by
men with clubs and dogs to reach where the nets had been
laid. It was extremely dangerous work. The trappers had
connections to Ayia Napa drug gangs and violence was
commonplace. Everybody I was with had at some point
been captured and subjected to a nasty beating and they
feared sooner or later one of them might end up dead. After
scuttling on my hands and knees through a ditch for an
hour or so I had the first sighting of one of these traps,
spiralling out across a stubble field with artificial birdsong
booming out. It was a dismal example of how humans seek
to subvert such a natural wonder as migration and harvest its
spoils to our own ends.

At Spurn the modus operandi for catching the birds
was actually very similar to the Cypriot trapping gangs,
but here they were gently freed and taken to the ringing
laboratory. A gruff cockney called Paul showed me the
technique, the tiny songbirds absurdly delicate in his
meaty hands. That morning he had already ringed three
tree sparrows, a goldcrest, a sedge warbler and a blackcap.
Next he took hold of the tiny treecreeper, its breast snowy

white and the claws that enable it to pirouette up tree trunks almost absurdly curved in proportion to the rest of its body.

After measuring its wings and weighing it – a tiny 7.9g, which is light even by a treecreeper's reckoning – Paul set the bird free. Then it was my turn: a tree sparrow retrieved from the nets. Paul taught me how to ring, using a pair of pliers carefully to shape a metal band around its foot. As I worked I was desperate not to catch the sparrow's tiny claws. After recording its measurement, and still keeping a careful hold, I leant my arm out the window to set it free. The sparrow's vibrating primary feathers grazed my palms as it launched itself deep into the shelter of a hawthorn a few metres away.

Later, I sat out for an hour or so with the volunteers at the Spurn sea hide. It was a calm morning and we arranged ourselves in a long line with the waves rolling in around us and wind turbines spinning noiselessly across the horizon. The voices of the volunteers – many of whom had made their own migration to be here – mingled gently into the sounds of the sea as they called out the names of the species flying overhead: red-throated diver, brent goose, birds of the north country, two fulmars, another swift that should really be somewhere else.

A few seal heads popped up in the water as I stared out to sea where minke whales were also currently migrating along the east coast. Just south of us along the Humber estuary, the seals of the huge colony at Donna Nook were beginning to haul themselves out to breed upon the sand.

Sitting here on this tiny spit of land I was filled with a sense of the world on the move. Near to us along the beach were the ruins of a 12-inch gun placement built in the Second World War that fell into the sea twenty years ago. The hide behind us would also soon need to be moved back in order to save it from being swallowed up by the encroaching tides. Out here on the very edge I became acutely aware of the great movements stirred up by the seasons – of life in constant flux. I thought of a passage in Edward Grey's *The Charm of Birds* when he reflects on the difference in birdsong between spring and autumn: 'Is it the song or I that have changed?'

CHAPTER SEVEN
Budburst

In the dying throes of the winter of 1913, the writer and poet Edward Thomas set out from his London flat on his bicycle heading towards the West Country. His loosely appointed finish was the Quantock Hills, but his principal objective as he passed through the city's rain-lashed streets was to travel in pursuit of spring. Thomas headed west through suburbs named, he noted (in the same manner as modern developments today), after the very pastoral visions of countryside they were rapidly destroying. Near Wimbledon, he encountered a roadside hawker selling songbirds in a cage: chaffinch, scruffy for a shilling and those of a neater plumage for 18 pence; goldfinch at five shillings and sixpence; and linnets at half a crown rattling against 6-inch cages that reminded Thomas of a beating heart.

The poet wrote of a journey fuelled by nostalgia. He described Easters of his youth in which he could recall seeing swallows on Good Friday, hearing cuckoos the following Saturday and nightingales singing by Easter Sunday. Thomas provided an early description of what would become a national affliction over the course of the twentieth century: a growing sense of somehow being unplugged from the seasons. In London, Thomas lamented, many days have simply 'no weather'. Instead the four seasons are reduced to merely hot, dry, cold or wet.

Thomas did eventually catch up with the weather of his memory. He encountered the first flowering hawthorn of the year in a churchyard in Glastonbury and, later on, bluebells and cowslips marking what he called 'the grave of winter' from which blossoms spring and the season is finally vanquished. Upon reaching Cothelstone Hill in the Quantocks on 28 March, Thomas described completing his quest, vaulting straight over April and into the month of May. Thomas's book, *In Pursuit of Spring*, was published in 1914, the year in which the Great War broke out. Three years later the 39-year-old was to meet his own journey's end. After enlisting as a private in the Artists' Rifles he was killed in 1917 during the Battle of Arras.

In the dying throes of the winter of 2019, Liz and I set out from our Sheffield home to walk a section of the Pennine Way. We had been attempting to complete the 270-mile route in stages over several years and had done so already in strange weathers. The Pennine Way leg we decided to undertake on February 2019 was our first long walk of the year. We didn't plan it especially with Thomas's book in mind, but I did consider, as we rattled off on the train from Sheffield station, how our planned walk that day was our own journey in pursuit of spring.

Liz had bought me a copy of Thomas's book for my twenty-eighth birthday in 2012. I still keep the accompanying card she made for me: a hand-drawn landscape of a view of Swaledale (from Reeth to Langthwaite in the northern Yorkshire Dales), where we had walked in January of that year. The weather that weekend was memorable: beautifully crisp, with the bracken iced like spiderwebs and frozen earth resisting our footprints. We walked like ghosts over a ridge of scree slopes and crags known as Fremington Edge.

In February 2019 we set out for a different part of the Yorkshire Dales. The date made it a full month before Thomas's own journey started on Good Friday. In any normal year we would have expected to encounter few signs of spring, and more likely be in pursuit of the sort of hail showers that once left us – in another part of the Dales walking another leg of the Pennine Way – cowering alongside the sheep sheltering in the lee of the nearest drystone wall.

But 2019 was no normal year. By the twentieth day of the month the Royal Horticultural Society had declared it 'Fragrant February', as a spell of unseasonably hot weather (off the back of a mild winter and a broiling hot summer the previous year) had led to an early explosion of blooms. Viburnum, magnolia, witch hazel and honeysuckle had already burst into life and the air was heavy with the scent of their blossom.

That same heady, listless month marked Scotland's warmest February day on record, with the 18.3°C in Aboyne, Aberdeenshire, beating the previous high, which was measured in 1897. It was also the first time in February that the British Isles recorded a temperature over 20°C (20.3°C recorded in Ceredigion, West Wales). There was no real need for anybody to travel in pursuit of spring; it was prematurely careering all about us.

We set off on a Saturday to follow the fifth stage of the Pennine Way. We took a train to Keighley and then an empty bus to a rural stop on the edge of a non-descript

B-road which, according to the map, denoted the beginning of our route. Our walk that day stretched for 17 miles into the southern Yorkshire Dales. As we headed over the sheep pastures that spanned the limestone plateau we spotted wild primrose, snowdrops and yellow stars of celandine scattered across the grass banks. Over our heads great flocks of lapwing and curlew, newly arrived from their coastal wintering grounds, scythed through the perfect blue of the sky. The skylark were in full song and the shared refrain as we walked through a spring that should have been a whole month ahead of us was: 'This is not normal ... this is not normal ...'

The Yorkshire Dales is a landscape we both know well and have experienced in all weathers. It is where we would disappear for weekends away when we lived in nearby Leeds as students and during our early twenties. My first job as a young reporter at the *Yorkshire Post* was based in the Harrogate district office covering the Yorkshire Dales. We have whiled away many happy days here, walking or cycling or holed up in some quiet pub, talking of all that lay before us and the life we envisaged for ourselves. It is where one Monday morning we delayed returning to our jobs in Leeds and instead bunked off to walk in the softly falling snow of the grounds surrounding Jervaulx Abbey. It is the place where I caught the worst bout of flu of my life on an ill-advised winter cycling trip. It is also the place where we said our marriage vows.

Landscapes like this that are laced with meaning conjure curious feelings of nostalgia when we return to them. We talked as we walked that balmy February day of the proper winter we had experienced seven years previously walking over Fremington Edge and the conflicting emotions of relief and disquiet elicited by this unseasonal burst of warmth. And we talked also of our own personal uncertainty that now lay ahead of us.

When you struggle to conceive, you envisage milestones in the year which turn out to be false. You look ahead to birthdays, holidays and weddings and imagine being there with your new baby listening to the adoring coos of friends and family. Then the months pass and whatever event it was that lay ahead of you is now suddenly there, and nothing has changed. Except it has.

'Time passes,' wrote Dylan Thomas in *Under Milk Wood*. Even if your own internal rhythms are stalled, all life goes on.

The pain of infertility − already adept at surging from a barely there ache to something that literally lodges in your throat − can suddenly sharpen on meeting up with loved ones whose own families are growing. And then there are those who you know even without asking are enduring similar difficulties.

You can share their sorrow, and feel your own, all at the same time as being delighted for another friend celebrating a pregnancy or new child. It is possible to experience all these emotions at once and yet often impossible to convey them to others. How to explain to somebody you love that you still want to fully be a part of their happiness, even if it can temporarily crush you under its weight?

The steady march of the seasons, as with the lives of others, serves as a constant reminder that life does not pause even as our attempts to create new life remain unfulfilled. The year speeds by like a train. Suddenly the clocks have gone forward again. Time passes.

★ ★ ★

Over the decade that I have written my *Telegraph* weather column, I have kept a weather diary of sorts; a note of the seasons as they change about me and potential ideas to write about each Saturday. During these difficult recent years, that weather diary has helped root me to life outside of my own.

Every year I note the first flowers, and frosts, and bird arrivals and departures in my garden. The first tadpoles in our pond in spring, and the moment each autumn where the downy birch at the end of my garden seems to catch alight and then shiver down all its leaves at once.

We keep chickens at the back of the garden and I also keep note of their seasonal shifts. We have had assorted arrangements over the years. Largely they are rescue hens (although once we bought some fancy birds from a farm on the edge of Sheffield) and the greatest satisfaction is watching the ex-battery birds come to life.

When they emerge, timid and bald from their cardboard box, the rescue hens have never experienced anything beyond the 24 hours of artificial light they are subjected to in the battery farms to force them to lay as many eggs as possible. Often the combs on their heads are distorted as the birds have to grow them larger in an attempt to dissipate the stifling heat.

The weak, pale birds are the extreme embodiment of a life lived outside of the weather. They have never felt the sunshine on their feathers or had any concept of rain. Plus their claws are several inches too long, because wedged together in their pens they cannot fulfil their innate desire to rake the earth for grubs.

Over time, we watch the birds become weathered. They learn how to dart for cover when it rains and on sunny days bask in the dust bath we have made for them, kicking soil up into their feathers in a cascade of filth. As they do so, their eyes literally glaze over with pleasure and they emit a curious burbling sound which, I assume, is a sort of poultry purr. After months, sometimes even weeks, of exposure to this new life outdoors, their combs and wattles shrink to normal size and turn a glorious shade of red.

Over winter they shed their feathers in order to grow new plumage. At times they can become so bald you see the barbules poking out of the pockmarked pink skin. During winter they typically stop laying eggs for several months to conserve their energy to cope through the colder nights and shorter days, although for the first time during the winter of 2019/20 – a winter without snow – I noted in my weather diary that they kept laying all the way through.

In the ninth century, Pope Nicholas decreed every church in Europe required a cockerel on its steeple or dome as a symbol of Peter's denial of Jesus – most added a chicken to an existing weather vane for sake of ease. The oldest remaining example of this is the Gallo di Ramperto in Brescia, commissioned around the year 820. Chickens persist in the form of weather vanes across the world. Watching the reaction of our hens to the weather, I often feel as though I am experiencing the seasons through their own beady, unblinking eyes.

What my hastily scrawled (and largely illegible to all but me) weather notes combine to provide is a rough picture of how the seasons are affecting life in our garden in any given year. They help me spot patterns, too. Taken together over several years these diaries provide reassurance that certain things will happen at certain times. The first chiffchaff of the year singing in the woods behind my house, for example, has always happened in the third week of March for every year I have lived here. Some things never change. Edward Thomas, for instance, described hearing a chiffchaff on 19 March when he was 15 years old, and not a year passed without him hearing the bird within a day or two of that date. Each note of the bird's song, he once wrote, was a tiny nail hammered into winter's coffin. But my diaries also reveal where the weather is doing strange things and nature veering spectacularly off course.

That same record-breaking February in 2019, I logged a wasp knocking at my study window on the eleventh day of the month. I recorded strange budburst and dandelions in my garden when normally they do not appear until the following month. The hottest day was on Monday, 25 February, which started with a hard frost covering my garden and ended with me watching a cabbage white fluttering along the drooped heads of some flowering hellebores.

Another winter had suddenly passed and we were emerging into this premature abundance of spring fecundity back where we had been the previous year. I wrote, then, that the grass is growing greener by the day. And I wished for a moment to jam a crowbar in the axles; to slow the whole spinning earth in order to allow us to catch up.

* * *

In the 1950s, a British naturalist with a burgeoning literary reputation embarked on a phenological study of his own back garden. Richard Fitter, who in 1952 had published the *Collins Pocket Guide to British Birds* and was working on a similar guide to wild flowers, decided to start recording the flowering times of the flora in the garden of his home high up in the beech woods of the Chilterns.

A scientist by training who during the Second World War served in the Operational Research Section of Coastal Command, in peacetime Fitter had become secretary to the Wildlife Conservation Committee of the Ministry of Town and Country Planning, specifically helping identify new nature reserves. Fitter also embarked on a career as a journalist, becoming assistant editor of the *Countryman* magazine and in 1958 taking an enviable position described as 'open-air correspondent' for the *Observer*.

In 1954 he started recording the flowering dates of the species in his back garden. Over the next forty-seven years Fitter recorded 557 plant species in an exercise that would become one of the most authoritative phenological studies conducted in Britain, one which demonstrated the extent to which the changing weather was altering the world around us.

His study, published in 2002 and co-authored with his son Alistair, found that the average first flowering date of 385 British plant species had advanced by 4.5 days during the 1990s, compared with the previous four decades, while 150 to 200 species had started flowering on average fifteen days earlier in Britain. As well as proving temperature to be a key determinant of flowering time, Fitter claimed the data showed 'a powerful climate-warming signal' had become established during the 1990s, at the time the warmest decade on record. The ecosystem and evolutionary consequences, he concluded, would be profound.

Over the twenty years since that study was published, the warming trend Fitter identified has massively ramped up. As Fitter noted in his 2002 paper, it is difficult to detect biological patterns in the majority of phenological datasets as they are, by their very nature, studies conducted on a small local level. Much like early weather forecasts they depend to varying degrees on myopia, fixating on what is happening in our own immediate vicinity, such as the patch of hairy bittercress which Fitter found to be flowering nearly a month earlier over the decades he monitored the plant in his garden. But added together these studies create a tapestry of change occurring across the country.

One such study, published in 2015 by researchers at Coventry University, analysed more than 20,000 records dating back to 1891 of things such as oak leaves, hawthorn flowers, frogspawn, and the first swallows being sighted. Between the late-nineteenth century and 1947, the study concluded, spring moved up the country at around 1.2mph, meaning that it travelled around 28 miles per day; it would have taken nearly three weeks for the whole country to be in full spring. But according to the study now, the season speeds up Britain at 1.9mph, covering a distance of 45 miles each day. Spring, therefore, can be considered fully 'sprung' a full eleven days earlier than in the nineteenth century. Thomas Hardy wrote in 1917 of a 'backward spring', with primrose and myrtle bushes forcing their way through the frozen earth. These days, the season arrives on rocket boosters.

★ ★ ★

On Saturday 4 January 2020, the first few days of a new decade, I head over to North Yorkshire to take part in one such long-ranging phenological study. In 2012, the Botanical Society of Britain and Ireland (BSBI) launched its New Year Plant Hunt, encouraging people to record over a four-day period the species of plants in flower that they found emerging in the winter gloom. What was originally intended as little more than an excuse for a walk has turned into a burgeoning scientific survey with each year nearly two thousand participants recording plants in eight hundred locations across the country. While even a decade's worth of data is not yet enough to provide a baseline to demonstrate proper systematic change, the plant hunt does reveal how species are responding to changing weather patterns as a consequence of climate change.

Our meeting point is a farmhouse in a village close to Harrogate called Staveley, which is home to a wonderful nature reserve where I used to come and watch dragonflies in the summer when I worked for the *Yorkshire Post*. This village was also where I wrote one of my early scoops for the paper, about a small dead-end road – named Bedlam Lane – next to a field where a council highways boss kept some horses. One year, to the dismay of local residents, Bedlam Lane had been bizarrely included in expensive resurfacing works despite there being other, far busier roads in seemingly much greater need of repair. All impropriety was strongly denied by the council, but the paper included the story on their front page in any case.

The group of plant hunters I am meeting already know I am a journalist. According to the map, the farmhouse which marks our rendezvous point is next to a horse paddock and a fleeting fear crosses my mind as I drive up the A1 that perhaps this council highways boss might live there and somehow recognise my name (even if it was a good ten years on from the story being published).

I knock slightly hesitantly, and fortunately my paranoia is misplaced. Instead a woman called Muff Upsall opens the door and welcomes me into a busy kitchen where a dozen or so volunteer botanists are munching Christmas cake and mince pies and discussing the route we are to take on our plant hunt.

Over tea Muff tells me she is a retired teacher who had moved to the house eight years ago from Shropshire to be closer to her grandchildren. She chose the house not because of the adjoining horse paddock but its proximity to Staveley Nature Reserve. Muff says she had been badly injured in a cycling accident in 2003 and left unable to ride horses or walk long distances. Instead she watches and records: the plant species around her land, the owls floating over from the reserve and more generally the weather. During her time in Staveley, Muff tells me she too has kept annual records, and they show things getting wetter every year.

Our walk that day is being led by Kevin Walker, head of science at BSBI who lives in nearby Harrogate. At the same time as we are exploring Staveley, plant hunters are joining in from Shetland to Guernsey, from Donegal to Anglesey, and in cities such as London, Newcastle, Bristol and Galway. The earliest record submitted for 2020, Kevin says, was of someone scouring the roadside verges by torchlight on their way back from the New Year's Eve firework display over the River Thames.

Before heading into the nature reserve, we explore the field surrounding Muff's farmhouse. Kevin's two young children scamper off hunting for owl pellets and (playfully) hitting each other with sticks while I talk with Kevin and his partner, Claire, a grassland specialist. By the horse paddock we spot common speedwell, groundsel, petty spurge, and tiny white constellations of shepherd's purse that are already in flower. So too, dandelion and white dead-nettle.

Every year, Kevin tells me, the top four flowering species are daisy, dandelion, groundsel and annual meadow grass. Even though the botany books count them as spring flowers, Kevin insists to me that many have historically got it all wrong. 'We have this idea of when things should be in flower but that is actually the case with very few species,' he says. His explanation for why so many botany books incorrectly identify winter flowering times is simple: even dedicated scientists don't like going out in the rain.

Aside from correcting historical inaccuracies, the plant hunt is primarily focused on recording recent changes. Kevin says that around half of the flowering species they log are not things that have come into bloom early, but rather have persisted over winter due to mild temperatures not killing them off. Among these common so-called 'autumn stragglers' are giant hogweed, yarrow and ragwort.

This matters to the long-term health of the plants, Kevin tells me, as by remaining in flower they cannot build up their reserves for the following year. I think of these stragglers as indicators of how the four seasons are becoming disjointed. With no winter frost to force them into a period of proper dormancy the plants linger on, drawing on whatever energy reserves they can muster and persisting wearily into spring.

We walk towards the Staveley Nature Reserve and the guns that have been blasting all morning from some nearby New Year pheasant shoot grow louder. I am looking up at a copse of hazel and willow trees slung with silvery grey catkins like Christmas decorations when Kevin yelps in excitement that he has found a premature flowering rarity. In the shade of a goat willow tree he points out a patch of dog's mercury, whose small green studs of flowers have emerged but in fact should not normally be seen until spring. As I inspect the plant, Kevin adds with a botanist's glee that it is extremely poisonous.

Other so-called 'spring specialists' are also now being regularly spotted on the various New Year Plant Hunts. Cow parsley, daffodils and bluebells are all now a common winter occurrence. Kevin tells me bluebells are flowering on average roughly two to three weeks earlier than they were thirty years ago. Generally, the plant hunts are recording in excess of six hundred flowering species each year. A few weeks after our search around Staveley, Kevin informs me that 615 different species in flower have been recorded across the country during the 2020 hunt.

While the ten years of data gathered so far is not yet enough to demonstrate the effects of long-term climatic shifts on the flora of Britain, the plant hunts have amassed a treasure trove of evidence highlighting the way in which the weather trends of any given winter can impact upon the flowering times of plants. In 2016, following the warmest December on record in Britain with average temperatures of 4.1°C to 7.9°C above the long-term average, and one of the wettest ever with nearly twice as much rainfall recorded as normal, volunteers registered 612 flowering species. Most surprising of all that year were seventeen different recordings of hawthorn in bloom.

Steeped in folklore and known as the May-tree, the hawthorn is the only British tree named after its usual flowering month. There is historical precedent for hawthorn coming into bloom over winter. Before its destruction some years back, the Glastonbury Thorn was the most famous of its species in Britain. Said to have grown two thousand years ago from a staff driven into the ground by Joseph of Arimathea, the Glastonbury Thorn 'miraculously' flowered twice each year, at Christmas and Easter. The winter blooms were deemed so rare they were sent to the Queen to decorate her breakfast table on Christmas Day.

Typically, however, the hawthorn is a tree associated with May Day garlands. Its foaming white late-spring blossom was once used to decorate houses, symbolising the beginning of new life and providing a conduit to other realms. The thirteenth-century Scottish mystic and poet Thomas the Rhymer claimed that he once met a faery queen by a hawthorn bush from which a cuckoo was calling, and was spirited away for seven years.

In a nod to tradition – and mindful of the fact individual species can support in excess of three hundred species of insects – a few winters ago I planted a hawthorn outside our house. So far it has flowered as a hawthorn should in early May, though what I have seen from the plant hunt and elsewhere is how rapidly these days the abnormal can become normality.

It is not only our folklore and cultural assumptions of the seasons that are struggling to keep up with the weather. Botanists fear the ability of plants that have survived for thousands of years on these islands to cope with the current pace of change. While rising temperatures are driving insects, birds and butterflies north, for plants it is a different story. Kevin says that most species in this country are long-lived and cannot adapt quickly enough to the rapidly changing weather patterns they are currently being forced to endure.

There is an exception to this rule, however, now dotted across Staveley Nature Reserve and increasingly all over the north of England, which he is keen to show me. Orchids are different to most plant species in that they can migrate. Their tiny seeds waft up into the stratosphere and can drift for hundreds of miles. Since the 1980s, bee orchids, once restricted to southern habitats, have spread up the country and a decade ago were first spotted in Scotland. Southern marsh orchids, whose miniscule seeds weigh just 0.1 milligram, first appeared in the fields around Staveley a few years ago. Kevin says he has recorded the species in twenty sites around Harrogate.

In Staveley now there are also bee orchids, which come into bloom each summer, their pink sepals spreading like wings to reveal a velvet-textured lip intricately marbled with yellow, red and brown. The migratory instincts of the flower have even exceeded the insect it evolved to entice. The species of bee this elaborate display is intended to attract does not exist in Britain but the orchids still manage to self-pollinate.

This vibrant display remains a summer occurrence but Kevin is still keen to show me where the bee orchids are growing. He takes me to a corner of the reserve he wishes me to keep secret, for fear of hordes of orchid photographers treading all over the grass, and points out a few green rosettes which in a few months will transform into the sort of exotic flower once alien to this part of North Yorkshire but now becoming increasingly common.

This close examination of the ground is something quite new to me. Generally when I am outdoors my eyes are naturally drawn to the trees, in that involuntary upwards trajectory of a birdwatcher's gaze. Kevin catches me at it several times during the plant hunt and playfully urges me to keep my eyes down. In doing so we record changes that otherwise I would not have noticed occurring under our feet, of spring coming into bloom before the twelfth night is even upon us and the first emergence of exotic flowers in a field in North Yorkshire.

But even botanists still finish the day with birds. Later, before leaving the reserve, we head to the hides looking out over the lake in the hope of catching a glimpse of the water rails that live here – a species I have never previously seen. The water rails must have heard us coming and kept out of sight, but as the sun begins to set a barn owl lifts off, swooping silently over the reedbeds. I raise my binoculars to intercept its flight-line and as I focus on its ghost-white form, an imperceptible rustle catches its eye. The owl dives down, emerging with a vole in its claws.

It is getting cold now and my fingertips are numb holding my binoculars up. The reedbeds have turned to copper; shades of purple ink vein the sky. As the owl feasts out of sight on its evening meal, I think back to my Ladybird *What to Look For* series and how this snapshot reminds me of a page in the winter book, where the bulrushes stand like silver spears against the dusk. But I know this is only a brief glimpse of winter proper as I understand it in culture and memory, viewed through the narrowest of lenses. Outside of my immediate field of vision, in the flowering trees in my periphery and the premature blooms underfoot, the season is in disarray.

CHAPTER EIGHT

Winter Sleep

Spring 2020 washed in fast on the back of the wettest February on record. That winter had also been recorded as Europe's warmest ever: 1.4°C above the previous high, during that of 2015/16. Centuries of tradition melted away with the snow. Germany's famous ice wine harvest failed for the first time as none of its thirteen growing regions recorded the sub-zero temperatures necessary for the grapes to freeze on the vines to create the digestif. In Scotland, the owners of a sled-dog training centre in the Cairngorms announced its permanent closure due to the lack of snow – the first British business, it was said at the time, to close as a direct result of climate change. In Sweden, snow was shipped in for winter sports events, and from the boreal forests of the far north to the enclosure at Moscow Zoo, bears came out of hibernation an entire month earlier than usual.

All through the winter I had noticed blackbirds proliferating in my garden, gathering in flocks of six or seven at a time, feasting on the earthworms in the damp, frost-free soil. Blackthorn exploded into bloom early in the hedgerows. By February the frogs had spawned and sparrows started nest building in the cracked old sand and cement render covering the top floor of my house. Every day I watched the birds flash past my study window as they constructed their nest. Occasionally the odd downy white feather drifted past, reminding me of the snow I hadn't seen even once that winter – bar a day or two of slushy Sheffield hail which could sustain little more than a few seconds of footprint before dissolving into the road salt.

Watching the world coming so prematurely into life left me with a slightly sickly sensation. Philip Larkin once wrote of the trees coming into leaf marking the celebration that another year is dead and the world beginning afresh. But in these broken records on my computer screen and broken patterns outside my window, I started to sense not a seasonal cycle, but a spiral.

During the course of that winter I had been in touch with a man called Nigel Hand, amphibian and reptile expert and one of Britain's leading authorities on adders. Typically between October and March each year the snakes go into a period of dormancy known as brumation – a term first proposed by the US zoologist Wilbur M. Mayhew and derived from the Latin *bruma*, by which the winter solstice was known in the Roman Calendar. After consuming their final meal around September, the snakes typically retreat underground before rousing themselves in spring to bask, battle and eventually breed.

Nigel has been monitoring adders, and his own local patch in the Malvern Hills, since the 1980s, but in recent years has started noticing the snakes emerging at curious times. His earliest sighting came on 28 January 2018, and he has now received reliable reports of adders spotted in every calendar month in Britain. The snakes have even been seen on Christmas Day and Boxing Day.

In November he emails me a photograph of a female adder cruising along a grassy path not far from his home and later that month sends further news of another sighting of a large female adder on the Mendips. Rising temperatures aside, winter floods are also driving the snakes out early as their hibernacula can quickly become deluged.

While out reporting on the February floods of the River Severn, I emailed Nigel a photograph of a washed-out snake collected by a fire and rescue team in the town of Bewdley. 'Grass snake,' he replied, and told me there had been another sighting of one that same week at a flooded rugby pitch in Droitwich.

After several failed attempts to meet, when the record rainfall disrupts our plans along with the snakes, a brief respite of high pressure looms and I drive to meet Nigel on the first Monday morning of March. Typically this would be when the adders, Britain's only venomous snakes, are just beginning to emerge, but Nigel tells me that he now sees them in early February in his patch on the Worcestershire–Herefordshire borders.

Searching for snakes is unlike the type of wildlife watching I am used to. Indeed, prior to meeting Nigel I had never seen an adder in the wild. I arrive at his home, an old brick-built railway worker's house on the edge of Ledbury, around 9.30am and expect us to head straight up into the hills. But instead he puts the kettle on and encourages me to sit in an armchair.

To see snakes, I slowly start to understand after a while in his company, we must think like snakes: ectothermic animals whose regulation of body temperature depends on the world around them. Male adders emerge first each year and face a daily struggle to raise their body heat to between

25°C and 30°C in order to ripen their sperm and develop a suitably strong condition to mate. And so, while we wait for the world outside to warm, we bask in our armchairs, sip tea and talk snakes.

That is a subject difficult to avoid in Nigel's front room. In a bowl on the dresser are freezer bags filled with adder skins he has collected for genetic research – the snakes typically first slough their skins in the middle of April, although again are now doing this earlier in the year. On the table are the radio telemetry kits he uses to record the movements of the snakes, and on the rug in front of the wood burner is a chainsaw for pollarding trees and creating suitable habitats.

Despite his efforts, locally the snakes are in rapid decline and nationally in crisis. Across the Malverns, he tells me, there are just three major populations left while in the nearby Wyre Forest – monitored by the naturalist Sylvia Sheldon (a satisfyingly alliterative name for a snake expert) – a healthy population of three hundred or so adders recorded in the 1980s has plummeted to thirty-nine today.

The causes of the decline are depressingly familiar: habitat loss, pressure from a lack of genetic variability and climate change. The warmer wetter winters of recent years are, Nigel says, of real concern and he fears the weather could be behind recent evidence of fungal disease among snakes in the wild. Also, when the adders come out of brumation too early it means they are vulnerable to sudden cold snaps. He shows me a paper he has written on a gravid (pregnant) female adder he encountered in the Malverns on Valentine's Day, 2013. Adders normally mate in late April or May and give birth late August or early September that same year. The previous year, 2012, had witnessed a particularly wet and cold summer so Nigel presumed she had delayed giving birth until the weather improved.

Over the course of that spring he continued to watch the female basking in the sun but at the end of March, the weather turned, heavy snow fell on the hills and the temperature dropped to -3°C. After several failed attempts to locate her he discovered her body on 7 April, under a bed of bracken litter in which she had clearly burrowed in an attempt to get warm. Nigel conducted a post-mortem examination and discovered ten fully developed young snakes inside her – all dead.

★ ★ ★

Driving with Nigel towards his patch I learn a new adjective: 'addery'. A landscape, it transpires, can be addery, a smell can also be addery (Nigel says the scent of the snakes reminds him of wet hessian bags). And as we talk I decide that a person can be addery, too. For Nigel is clearly enchanted by the snakes, entwined with their lives.

It started as a boy, growing up in the Black Country, when he discovered a small population of adders in a wild edge of the school playing fields. His science teacher didn't believe him at first, until he showed him their hideaway. It proved an early lesson, too, in conflict with the snakes. He recalls a couple of school kids throwing a snake between them, a biting incident, and eventually the adder's meagre habitat going up in flames, torched by a cigarette lighter or discarded fag butt.

That experience did not deter Nigel. He was, metaphorically, bitten. 'I just had a knack of finding them,' he says. Even when his dad made him caddy for his usual round of golf he would be forever loitering in the rough, seeking out snakes, or inspecting the watercourses for great crested newts. These days his hearing isn't what it was, something Nigel blames on attending too many punk gigs during the 1970s and 1980s. His other senses, though, remain tuned with a reptilian keenness.

I'm not allowed to say exactly where we are, for fear of the adders becoming overrun, but the snake habitat he has helped develop, along with a group of local volunteers, is from a distance just a scrubby hillside. Look closer, though, and they have carved glades for the snakes to glide through the bracken and built flat terraces into the slope, what Nigel calls 'adder patios'. It is a beautiful, clear morning but the weather havoc of recent days – and months – is not difficult to see. In the distance the inundated floodplains of the River Severn sparkle under the blue sky.

It does not take long to see our first adder. In all likelihood I would have hunted all day with no luck, but Nigel notes its presence immediately. The snake is about 15cm long with the muddy colouring to its dogtooth-patterned skin that denotes a male recently emerged out of brumation. Its head remains lodged in the bracken but its body is neatly coiled like rope on a ship's deck to best soak up the sun.

I watch as its scales enlarge and glisten with each exhalation. Suddenly, perhaps the result of one of us breathing too loudly or simply shifting weight from one foot to the other, the snake detects it is being watched and slithers back into its hide. The sound of it travelling over the dried bracken is a long singular note compared to the pitter patter of the common lizards we also see that day. It transpires noise can also be addery.

The snake secreted back in its hide, I stoop down and touch the brittle fronds of dried bracken upon which the adders bask. I trace my fingers across an entirely different climate populated by the snakes. The bracken itself is as warm as a cafe table on a broiling summer's day but I dig down just a centimetre or so into the cool heathland humus below. The bracken is necessary to the snakes but also poses a grave threat. During prolonged dry spells, Nigel says, the land here can quickly transform into a tinder box, leading to wildfires which are lethal for all life on the moor – particularly snakes.

Our attention draws away from the ground and I notice we are being watched. A buzzard perches at the top of a nearby oak, scanning the bracken for prey. The raptors target adders and with their populations increasing at the same time as those of the snakes are diminishing, Nigel is afraid they are being penned in. He is even more concerned about the squawking pheasants we hear nearby. A neighbouring shoot releases sixty thousand each year and he has come across evidence of the game birds devastating snake populations: chasing them from their burrows and pecking out their eyes. Considering the damage my own three chickens can do when left to roam our garden unchecked, I am unsurprised to hear of the destructive capacity of the pheasants on the landscape.

We walk higher up until we are short of breath and looking down on trees laced with balls of mistletoe. I hear a familiar croak and notice two ravens spiralling about us. Past more brambles and brash (the pruned branches of trees left lying on the ground that Nigel has created as habitat for the snakes) we spot glistening beads of cuckoo spit and the oily bodies of bloody-nose beetles, which, like everything else on the heath, seem to be out a month early. Nigel tells me the beetles take their name from the ruby droplets they exude when they feel under threat. He has handled them in the past and been left with red streaks on his skin.

An adder, too, has previously drawn blood. Only once, twelve years or so ago. At the time Nigel was working gruelling night shifts but rather than sleeping during the day he could not help but head out to his patch to monitor the adders. One morning, after one such sleepless night at work, he encountered an adder on a path and picked it up to assess its condition. He knows how to handle adders safely but his movements this time were too hasty and the snake reacted badly, coiling its body around and biting him on the knuckle of his left hand. 'I wasn't careful enough,' he notes.

At first he hoped it would be what he calls a 'dry bite' but soon started to notice a metallic taste in his mouth, which he likens to licking a battery, and took this as a definite sign he had been envenomated. It is a short drive from the adder site to his home and he made it before the next phase of the poisoning commenced. A deep rash bloomed on his arm, which started to swell up to the shoulder.

Soon he was suffering vomiting and diarrhoea. He cancelled that night's shift and in the afternoon received a phone call from his wife asking him to pick their sons up from school. By then Nigel's speech was starting to slur. At first she asked if he had been drinking – not, he is at pains to insist, a common daily occurrence – but then she guessed the answer: snake bite.

Adder bites are exceptionally rare, somewhere between forty and a hundred occur in Britain each year. The last death in this country was in 1975 when a five-year-old boy was bitten on the ankle in Scotland. Prior to that there had been just thirteen deaths in the previous hundred years. Having defended the snakes against nefarious scare stories in the media for years, Nigel was well aware of the figures and rather than head to his nearest accident and emergency department he took the decision to, in his words, 'sit it out'.

As the venom spread through his body he approached his condition with a biologist's zeal. He documented the various livid shades his skin turned and took notes of his other symptoms.

He was bitten on a Friday, and the following Wednesday, his condition improving though far from recovered, he finally visited his local GP, who immediately dispatched him to hospital with an overnight bag.

As news of the adder bite spread among staff, Nigel found himself at the centre of a queue of consultants and students jostling to assess his condition. In the end he had no need for antivenom and his body flushed itself of the poison. It took over a month, though, he says, for the mobility in his fingers to return. He still has the photographs of his injuries; an adder anthology of sorts. A reminder too, I suppose, of the lethal capacity of the snakes, and his own freely accepted folly at causing one to bear its fangs.

★ ★ ★

In 1900 an article was published in the *British Medical Journal* which claimed to reveal the discovery of human hibernation. The article focused on a group of Russian peasants in the Pskov region of the country who supposedly practised a custom 'akin to hibernation'. At the first snowfall of winter, the author wrote of his findings: 'The whole family gathers round the stove, lies down, ceases to wrestle with the problems of human existence, and quietly goes to sleep.'

Once each day the peasants rouse themselves to eat a piece of stale bread, a six-month supply of which has been baked in bulk that autumn, washed down with water. Family members take turns to tend the fire while the world around them sleeps under a soft blanket of snow. The author of the report coined a word for a custom he claimed passed down through the ages: 'Lotska'.

Germans call hibernation *Winterschlaf*, or winter sleep. It is a process that has long fascinated humans, not least because of our inability to replicate it. The Ancient Greek philosopher Aristotle believed the swallows which disappeared each winter were in fact hibernating in tree hollows. How our ancestors whose harsh lives were uninsulated from the ravages of the weather must have longed to curl up for the cold months to emerge renewed in spring.

But what of the awkward physical demands of the human body, which the author of the *BMJ* article failed to address? Adders, for example, are content to take their final meal in the autumn before sleeping through until spring. Our bodies, our appetites, deny us the ability to fully shut down.

Hibernation fascinates and tempts us, largely for the joyous simplicity of sleeping for half of each year. But also, perhaps, it is because the animals that can do so boast an intimate connection with the seasons tantalisingly beyond humanity's reach. The author of the *BMJ* article sums up this note of envy in his dispatch on the dozing peasants, describing their winter sleep as 'free from the stress of life, from the need to labour, from the multitudinous burdens, anxieties, and vexations of existence.'

The coronavirus pandemic brought about as close to a period of human hibernation as I have ever witnessed, when we collectively retreated back inside our burrows. In July 2020, when Prime Minister Boris Johnson announced the relaxation of the first lockdown, he declared it the beginning of the end of Britain's 'national hibernation'. In the early days of the pandemic, before the death toll started to increase and the virus inch uncomfortably nearer to those I loved, I admit to having felt a certain thrill. The air improved, city streets and waterways cleared of pollution, litter and traffic. As globalisation slackened the world seemed to expand once more, and as one of those fortunate enough to be in a home where I wanted to be during lockdown with a person I wanted to be with, my place in it felt more secure. Stepping back proved an appealing aspect of an otherwise frightening time. A brief period of waking sleep before it all kickstarted again.

During the 1950s, interest in the possibilities of human hibernation surged, in part funded by the US space agency NASA, which was keen to understand the potential physical limits of its astronauts. One recipient of that funding was the

British scientist James Lovelock, who at the time was experimenting on freezing hamsters. He would reanimate the rodents with a hot spoon pressed lightly to their chests, or with a homemade microwave gun constructed out of old radio parts, to restore their heartbeat before the rest of the body thawed.

While pioneering in its discoveries, Lovelock's work did not lead to a breakthrough on sending humans into hyper-sleep, although decades later surgeons would start using extreme cooling to induce patients into a period of hypothermia, slowing the body's processes in order to perform heart transplants. The NASA funding also allowed Lovelock to develop what would become his world-famous Gaia hypothesis – the belief that the earth and its life forms together constitute a single self-regulating organism, which has maintained itself for 3.8 billion years. After suggesting his theory of earth as a feedback system, Lovelock started to warn of its disruption by one dominant species; of the degradation of biodiversity and natural resources, and of the carbon we release into the atmosphere reflecting back at us through ever increasing global temperatures.

Lovelock has argued that humans are the consciousness of the planet, and yet by detaching ourselves from it we have created our own demise. It will take hundreds of thousands of years, he argues, but Gaia will eventually recover from the sixth mass extinction and temperature rises of 8°C and the collapse of the world as we know it. Humans, however, will no longer be around to see it. Our species will realise too late what we have gambled and lost.

Aside from the small-scale climate shifts such as the Little Ice Age, the Holocene era (the 11,000 years from the end of the last major glacial epoch to the modern day) has provided weather so remarkably stabilised that species have been able to evolve to be perfectly in tune with the seasons. Now, in a matter of decades, a process that has taken thousands of years to develop has started to unravel.

★ ★ ★

One of the most totemic of hibernators, in British culture at least, is the dormouse. Like generations of schoolchildren I grew up with the stories of Lewis Carroll's *Alice in Wonderland* shaping my formative imagination, in particular the somnolent rodent who curled up to sleep in the Mad Hatter's teapot and incoherently mumbled to the assembled guests.

In the Victorian era dormice were a favoured companion of schoolchildren, who used to stow the sleeping creatures in their pockets before releasing them after dark. Beatrix Potter also kept one as a pet, enchanted by a creature that spends two-thirds of its life asleep.

In the late 1900s, a naturalist called G.T. Rope published an article on the range of the dormouse in England and Wales, which described the species as present in all counties right up to the Scottish borders and 'common' in southern England. A century later new work was undertaken to assess how dormouse populations had altered.

Nocturnal and arboreal, slumberous and secretive, dormice are particularly difficult to spot. Instead populations are monitored through other signs, such as the hazelnut cases littering the woodland floor. Dormouse gnaw at nuts in a unique way, leaving a smooth inner ring and tooth marks spiralling around the outside. Attempts to map populations in the 1980s and 1990s discovered a disturbing lack of any such evidence of dormouse activity. It was eventually concluded that they had declined from 35 per cent of counties and had shrunk to a range that encompassed only southern England and the Welsh borders. In 2016 the dormouse was declared extinct in seventeen English counties, populations having plummeted by 72 per cent since 1993.

The spring before my adder hunt I made an appointment with a group of volunteers studying dormouse populations in Surrey to join them on a morning count of a woodland where the species continues to cling on. In doing so I wanted to understand the extent to which the changing weather was influencing their decline.

Dormice are a species that evolved to be beautifully in tune with the seasons. Successional feeders, in early spring they feed on the flowers of hawthorn and oak before progressing on to honeysuckle and bramble as they come into bloom. Over summer they alter their diets to encompass caterpillars, aphids and wasp galls before gorging on blackberries and hazelnuts in autumn. By October, once they have reached an ideal weight of around 30g, which allows them to endure winter hibernation where they lose a third of their body mass, they burrow down into leaf litter or at the base of hedgerows (their exact hibernating spots remain much of a mystery) where the fluctuations in temperature are less severe and then sleep out the winter until May, when suitable food is available once more.

Modern life has brought about significant disruption to this age-old routine. Nowadays, explains Julie Mottishaw, the co-founder of the Surrey Dormouse Group as we tramp through an old beech wood on the North Downs, the dormice are tending to wake up during warm periods over winter and head out from their nests in pursuit of food, only to find that none is available. We meet in the middle of April when dormice should typically still be in hibernation, but Julie says it is not uncommon these days to discover them still out in March and November.

Once they are awake for the year, dormice will weave delicate nests out of honeysuckle and bluebell stalks to sleep in during the day. The nests are also a sanctuary during cold snaps where dormice enter into a period of partial hibernation known as torpor, a state in which the processes of their bodies slow. Prolonged wet weather, in particular, is disastrous for dormice. Unlike other rodents they do not possess any oils in their fur and so water can soak straight through to their skin.

Julie's group has erected 1,200 boxes across twenty-four sites in Surrey for the dormice to build their nests, and regularly monitors them to see how early they have emerged from their hibernation and their breeding progress during the year. We head out to check them on a warm spring day. Chiffchaffs belt from the treetops whose new leaves are a brilliant emerald green, and electric yellow brimstone butterflies – among the first of the year to hatch – dance along the grass verges. We walk over a carpet of star-shaped wood anemone, flowering white in the dappled shade. The flower is an indicator of well-coppiced woodlands, the sort of habitat which dormice require and which has been rapidly eroded across Britain.

Julie tells me it was the conservationist Oliver Rackham who first pointed to the correlation between the decline in ancient woodland during the twentieth century and the loss of dormouse populations. The wealth of wildlife one encounters in such precious habitats makes a mockery, she says, of developers' claims that they will replant what they cut down.

Aside from when they are hibernating, dormice tend to avoid the ground. Instead, when they are active at night, they dance through the treetops in a delicate trapeze act, balancing themselves using their sensitive furry tails, which mark them out as unique among rodents in Britain. They are susceptible to owls and badgers on their nightly prowls but the main threat – aside from the loss of suitable habitat – remains the weather.

We fan out as a group and begin to open the boxes. A blue tit flies out of one, darting up into the boughs of an old oak tree to squeak its disapproval back down at us. In another we find a clutch of pale pink blue tit eggs resting in a beautifully woven nest of moss and feathers. We quickly close the box back up and leave it be.

We search dozens of boxes right across the woodland that morning but to no avail. Even this early in the season an unsuccessful dormouse search is a worrying sign as many of the creatures do not survive the rigours of winter. Dormice are unlike other rodents in that they are far from prodigious breeders. The dormouse follows a similar life cycle to humans. They live around six years and are slow to have young. If one litter dies, Julie says, then essentially a whole generation is lost.

A second site is suggested. It seems an unlikely location for dormice, a patch of scrubland close to a golf course, though activity has been previously reported there. We head over to search more empty boxes until we reach the final one, box 49 of the day, and a hushed cry of glee goes up.

Inside is a dormouse in a state of blissful repose. One of the volunteers who is skilled at handling the creatures gently lifts it out and tells me I can briefly hold it in my palm. It is weightless and curled into a ball, its long whiskers and chestnut tail twitching with each deep breath. Its tiny pink front paws are bunched and one pink foot is stretched akimbo. Just before putting it back into its box I lift my palm up to my ear and discover an ethereal sound – ever so gently, the dormouse is snoring.

CHAPTER NINE

Muirburn

The scars are growing fainter now, though a shadow remains scorched into the earth across Saddleworth Moor. On one side of the firebreak the bilberry bushes are fruiting and emerald green patches of sphagnum moss shine like spotlights on the moorland grass. On the other side, across the blackened peatland, the first shoots of heather are just beginning to emerge from the skeletal ruins of the worst English wildfire in living memory.

It is a year on from the Saddleworth Moor fire of 2018 and I am standing with Kate Hanley, the RSPB site manager for the moor, at the demarcation line denoting the point where it was finally brought under control. In total around 1,000 hectares of moorland were consumed in the wildfire, which broke out on 24 June and burnt for weeks. The smoke billowed across Manchester thirteen miles away, creating an ash cloud over the city which turned the sun red and temporarily lowered temperatures by 3°C. NASA satellite images confirmed the fire was visible from space.

The day itself Kate and I had been on different sides of the fire. She and her colleagues had been alerted the previous evening but waited until first light to go up on the moors as it was deemed too dangerous in the dark. Kate has worked up on Saddleworth for a decade and in that time has seen a lot of wildfires. The intimate knowledge of the landscape she and her colleagues have gathered meant her most useful role during the fire was in logistics on the ground, driving quad bikes between fire crews, delivering hoses and identifying new hotspots flaring up. She initially describes her memories of the day to me in a matter-of-fact manner as 'lots of flame, smoke and lots of worry'. Only after we have explored the scene together does she let on a little bit of the reality of what she faced.

The fire occurred during the peak of breeding season and she tells me about finding nests of the wading birds she and her colleagues have worked so hard to encourage to breed on the moor frazzled up to a crisp, whole generations lost to the flames. She personally had to put out a few burning hares she encountered. 'It's a shame, a real shame,' she says quietly as the wind howls over the moor. 'I remember at the time it just made me angry to see.'

At the same time as Kate was working alongside a loose coalition of more than a hundred firefighters, mountain rescue teams, soldiers, staff from RSPB and United Utilities, which own much of the moor, and gamekeepers on a neighbouring grouse shoot whose land was the epicentre of the fire, I was hotfooting it up from my newspaper offices in London along with the rest of the national press pack. I had been dispatched by my editors early that morning as soon as the first images appeared online and taken the train to

Stalybridge and then a taxi to the village of Carrbrook, which stands at the foot of the moor. I arrived around midday, by which time some fifty homes had been evacuated around the moor, including thirty-four in Carrbrook during the night.

Smoke billowed down off the moor, obscuring everything around us. The fog was thick enough to sting my eyes and public health officials were handing out surgical masks to residents and journalists to protect us from the worst effects of breathing in the toxic air. Conducting interviews in the masks felt like a novelty at the time, though it proved grim preparation for the coronavirus pandemic two years later.

Among those I spoke to was a bus driver who was evacuated after smoke began seeping in through his roof, which had been damaged only a few months ago by strong winds during the Beast from the East. He, like everybody else, was extremely worried about the damage the burning peat might be doing to his lungs. Peat fires are especially dangerous to health as they release carbon monoxide, benzene and hydrogen cyanide, which can travel hundreds of miles from the source. The upland bogs of the Peak District also hold two centuries of industrial pollution, once produced by the surrounding mill towns and large manufacturing cities such as Manchester and Sheffield. Stored in the peat for decades, as it burns those residues of toxins and heavy metals begin to seep out into the air.

As the fire worsened during the day, the concerns of residents grew more vocal. At some point in the afternoon the watch manager for Greater Manchester Fire And Rescue Service convened a hastily arranged press conference on a street corner in which he assured us they were taking air samples every hour, which were analysed by Public Health England. Should the smoke grow worse, he said, then they would introduce a full-scale evacuation of the whole area.

A photograph was circulating online that day of a woman walking down the high street in one of the surrounding villages wearing a Second World War gas mask and clutching her shopping bags from the Co-op, and I was tasked with finding her. As I drove up over the tops to the next village along, the full extent of the fire started to become clear. The valley below was blanketed in clouds of dark grey with only the old mill chimneys poking out. The local MP described the scene as 'looking like Mordor from *Lord of the Rings*'.

I filed my story for the following day's newspaper from the corner table of a hillside pub. As I typed, the regulars sat staring at the rolling news on the big-screen television normally reserved for sporting fixtures. The scene reminded me of a Hollywood disaster film where people watch events unfold while propping up the bar.

When I got home late that night my chest was rattling and my clothes smelt like a bonfire. I undressed in the kitchen, put everything immediately in the washing machine and stood in the dark for a while feeling the cool tiles on the soles of my feet. The firefighter crews and Kate and her colleagues remained up on the moors, and would be for another three weeks as the peat continued to burn.

★ ★ ★

That summer it seemed as if all of Europe was in flames. In Greece the government declared a state of emergency after wildfires destroyed the Aegean coastal resort of Mati, killing around a hundred people. Some rushed into the ocean to try to escape. In one villa twenty-six men, women and children were discovered dead, huddled in each other's arms.

In total 178,000 hectares of land was destroyed by wildfire across Europe in 2018. While that tally was actually less than the previous year, the shock was the number of countries involved. Sweden suffered the worst year in its history; Norway, Ukraine and Latvia were also badly affected. At one stage in mid-July, eleven separate wildfires raged inside the Arctic circle.

From Saddleworth Moor to the tundra of the far north, the blame was largely placed upon the weather. That summer records were being broken across the northern hemisphere after what had already been the second-hottest June on record. Britain was engulfed in its third-longest heatwave in history. As Saddleworth Moor was still burning, firefighters also spent weeks trying to put out a similar-sized blaze in Winter Hill in Lancashire, which at its peak covered an area of 7 square miles of moorland.

Aside from the more visible species such as curlew, dunlin, golden plover and hares, the moors sustain an intricate web of life. I think of the adders and common lizards I spotted with Nigel Hand secreted beneath the moorland bracken of Herefordshire, and understand better the sheer destructive capacity of a wildfire. On Saddleworth and Winter Hill entire ecosystems were wiped out in an instant. Experts said they would take years to recover, if indeed they ever could as such fires were now seemingly becoming an annual occurrence.

The weekend after Saddleworth Moor erupted I went walking with friends up in the Peak District. Even from twenty or so miles away we could still spot plumes of smoke from Saddleworth and Winter Hill, while helicopters whirred overhead carrying pails of water from the already parched reservoirs to dump on the burning moors in the absence of any rain. The sky was hazy blue tinged with the toxic fog rising up in the distance. The peatland underneath our feet was marbled with deep fissures and as crisp as a dried-out cowpat. We saw few signs of any moorland life as we walked. The very earth seemed to be crying out for rain.

Wildfires flickered in my thoughts until, a week or so later, an editor at the newspaper phoned, asking me to go to Portugal. The country was braced for a fresh wave of wildfires having in 2017 experienced its worst ever year, with 121 lives lost and thousands of homes and hectares of land destroyed in two major outbreaks in June and October. Following my experience at Saddleworth I wanted to understand why these blazes were continuing to increase in intensity and if there was any hope of ending the cycle.

At Lisbon airport I met a Portuguese journalist called Bruno Manteigas. Normally he lived in London as a correspondent for a Portuguese news agency, but we were to be working together on this story. He picked me up in his mother's Nissan Micra, which he had borrowed from her house nearby, and together we headed north for the mountains of Pedrógão Grande in central Portugal which had been most severely affected by the 2017 wildfires, where sixty-six people lost their lives.

Bruno was good company. He lived with his wife and daughters in the same bit of London where I grew up and was a keen wild swimmer. His great passion, though, I soon discovered as we stopped at a rundown motorway services a few miles north of Lisbon, was the Portuguese egg custard tart known as *pastel de nata*. Bruno was a true aficionado of the tarts and insisted on us buying and eating them wherever we could. He taught me the art of eating them: turning the tart over to inspect the spirals on the bottom, nibbling the pastry and then sucking up the custard which, he insisted, must not be oversweetened. Later that year, Bruno was planning to host a *pastel de nata* world championships in London.

If our journey up country initially had the excitable air of a road trip, that quickly changed when we turned off the main motorway and on to a highway called the N236-1. Suddenly the scale of the destruction was strikingly evident. Whole hillsides were covered in the blackened stumps of scorched forest. I discovered later that the N236-1 had been renamed by locals as 'the road of death'.

On 17 June, forty-seven people died here trapped in their cars as eucalyptus trees burnt through the valley, down to the very edge of the road, producing a heat so intense it melted the asphalt. In the nearby village of Nodeirinho, which was also engulfed, we interviewed residents who had sheltered together in a 3m x 2m concrete water tank as gas canisters exploded and houses were destroyed all around them. The worst sound of all was the animals: goats, ducks and chickens bleating and calling as they were consumed by the inferno.

One of those I met, a gardener in his early sixties called João Viola, had been among the motorists trapped on the N236-1 but managed to escape. He described driving through a 'rain of flames' and was just able to turn around and head back as the fire started to sweep down through the trees. Even as he drove out of the smoke he saw cars passing him in the other direction. He shouted and beeped his horn, frantically waving to stop them heading towards what he assumed was certain death.

After spending the night at an evacuation centre, he returned to Nodeirinho the following morning. He is proud of the village where he has lived all his life, and his home, which miraculously survived (next door burnt down). After giving me a tour of his garden where he keeps chickens and ducks in an enclosure round the back, we sat drinking coffee in his kitchen where he quietly explained the scene he was confronted by that morning and one he is condemned to continually relive in his thoughts.

Driving into the smouldering village, the first body he discovered was his cousin, Afonso, lying on the side of the road asphyxiated by the smoke. Of the eleven Nodeirinho residents who died that day, four were João's relatives. 'The worst thing for me was the silence,' he told me. 'No birds, no voices, no wind through the trees, just tears. Everything was black. It was like a bomb had gone off.'

I spoke to numerous forestry and climate experts while out in Portugal and they all seemed to agree on two things. Firstly, the wildfire season is lengthening. A few decades ago it was largely considered to begin around June and run through to late summer but that has now shifted to April and lasting into autumn, a pattern climate change will continue to exacerbate. The other was that these changing weather patterns are in themselves not solely responsible. The fires seem to exploit the very weaknesses we ourselves have created in the natural resilience of the countryside. Monocultures and intensively managed landscapes provide perfect fuel. The only solution, therefore, is mending our broken relationship with the land.

★ ★ ★

The longer we spent in Portugal the more parallels I saw with Saddleworth Moor. Both were indicative examples of long-term mismanagement of a landscape and how those past and present mistakes were being ruthlessly highlighted by wildfire. In Portugal, the obvious culprit was the vast eucalyptus plantations which cover the hillsides. Eucalyptus is known locally as the 'tree of fire' and played a central role in the tragedy that unfolded in the mountains of Pedrógão Grande.

Portugal is one of the most heavily forested countries in Europe, but also has the lowest amount of state-owned woodland – just 2 per cent. A complex system of inheritance, where land is divided up among relatives, has led to an impenetrable patchwork of ownership across the country, with 85 per cent of plots smaller than 3 hectares.

In previous generations this land was lovingly tended by smallholders grazing and growing crops, but as Portugal's population has moved to cities, many rural areas have been abandoned by the younger generation. For elderly family members left behind, the easiest cash crop to grow has become eucalyptus, which feeds Portugal's paper and pulp industry, worth billions to the country. Aside from the mammoth industrial plantation owners buying up tracts of countryside to transform into forestry, I was told numerous stories of local farmers being offered money to plant eucalyptus trees on their land and witnessed plenty of evidence for myself. Middlemen arrive at a farm and offer cash to temporarily take over the land and plant the trees which can be harvested in as little as seven years. It is an offer that for many struggling to make ends meet is too good to refuse.

The result has been a non-native tree, which originated in Australia, now covering an estimated 25 per cent of Portuguese woodland. The high oil content and deep roots of the slender eucalyptus make it adept at drying out the soil and preventing other native species from thriving. When set alight, the trees spread fire at an incredible intensity.

Britain's uplands have similarly seen much of their natural resistance to fire stripped away over centuries of mis-management which has led to them being left in a state of ecological crisis. Saddleworth Moor is a case in point: much of its peat-forming sphagnum moss has disappeared over the past century or so, and the wetlands have been drained, creating a monoculture of highly flammable heather – although in recent years there have been efforts led by Kate and her colleagues to restore blanket bog habitats on parts of the moor and similar work taking place elsewhere.

Large parts of the uplands have also been given over to the intensive rearing and shooting of red grouse, which has similarly increased its susceptibility to wildfire. In many instances gamekeepers seasonally burn large areas of moorland themselves as the tender shoots of young heather are deemed of benefit to the grouse. There are increasing steps to outlaw this, but each year when the burning season begins in October, smoke still rises from some of the grouse moors in the countryside near me.

This combination of draining, burning and overgrazing means the land is less able to withstand flooding and wildfire just at the very moment when, due to the carbon we continue to release into the atmosphere, those climatic pressures are rapidly increasing. I wonder about the 'wildfire season' as a term in itself and the scorch marks it leaves through each passing year.

★ ★ ★

In March 1965, a young Yorkshire archaeologist and poet called Jeffrey Radley published an article in *Nature*. Entitled 'Significance of Major Moorland Fires', it sought to examine the frequency and extent of wildfires on an area east of the Derbyshire Derwent between Penistone and Matlock, which encompasses the moorland above the west of Sheffield.

By the time of publication Radley was 30 years old and a prolific writer of archaeological papers. His main subject of interest was prehistoric archaeology and he made numerous discoveries out fieldwalking in the hills where he amassed a significant collection of flints (which following his death was donated by his wife to Museums Sheffield). In 1960, after a huge wildfire the previous year had cleared vegetation across Totley Moor on the edge of Sheffield, Radley discovered a Bronze Age shale-working floor containing twenty shale bracelet fragments.

He died young, and shockingly. In 1970 Radley was leading a team conducting excavations on the Anglian Tower in York, originally built by Anglo-Saxon kings. During one lunchtime break he climbed down into the excavated trench to inspect the morning's work and the supports gave way, burying him under the rubble.

A man fascinated by the layers of the past, Radley's *Nature* article sought to place the increasing number of wildfires which had started to be witnessed on the moors into some historical context. The aforementioned 1959 fire was, he said, the largest ever in the area, destroying hundreds of acres of moorland. As with 2018, the weather played its role. Derbyshire was the driest county in England that year and the period between May and September (when the fires broke out) was the driest since records began. Light anticyclonic winds meant Sheffield had received just 10 per cent of its monthly precipitation in September.

But Radley believes even against this backdrop of drought, it was people that started the fire. He claims the most likely cause was tourists who had gathered to marvel at the dried-up reservoir beds (just as I watched them do during the drought of 2018) and dropped lit cigarette ends, which ignited the tinderbox heather. The twentieth-century fashion for pre-rolled cigarettes in preference to pipe-smoking was, Radley noted, increasing fire risk.

Searching local records and newspaper archives, Radley compiled a list of the frequency of major wildfires, which he classified as those that had burned for two days or more, enough time for the heat to penetrate into the peat and destroy the rootzone of the vegetation the moorland sustains.

According to his research, from the 1920s a significant blaze seemed to occur once every five years or so; in the second half of the nineteenth century that rate had been roughly every twenty years. The earliest wildfire date he could trace was 1762 on Broomhead Moor, when several thousand acres were destroyed. Following that there was a long gap before the next, which occurred on the same site in 1826 when a bilberry picker reportedly emptied her pipe on the moors, causing several thousand acres to catch alight. The next fire after that was in 1868, engulfing the Hallam and Moscar Moors.

Radley drew a parallel between the increasing prevalence of wildfires and the parliamentary enclosures of the moors between 1760 and 1855. In the Peak District, as elsewhere, this state-sanctioned land grab replaced a communal system of utilising the moors with a different economy based on private grouse breeding – and shooting – and the exclusion of human and animal life in favour of game birds. A delicate balance of seasonal activity – grazing livestock and harvesting the produce of the moor – was thrown into disarray, causing scars in the land which stretch to the present day.

Prior to the parliamentary enclosures existed an archaic system of complex common and manorial rights that pre-dated the thirteenth century. Sheep were 'staff-herded' across specially designated sheep walks by flockmasters. Cattle owners were afforded rights to graze their livestock on common land through a practice known locally as 'stinting' or, pleasingly, 'summering'. As that name suggests, everything was organised in accordance with the seasons. Bracken, rushes, heather, timber, grass sods, bilberries and millstone were all carefully harvested and protected through manorial rights. In a separate paper Radley has researched the use of holly bushes as winter feed for livestock, which gives it fresh seasonal significance beyond that of Victorian Christmas pastiche.

Moor fires as a means of clearing the ground were, by and large, expressly forbidden. However, by the nineteenth century, as the enclosures had come into force and agricultural communities streamed off the moors to toil in the booming industries of Sheffield below, the gamekeepers took control of managing the landscape.

'Muirburn', writes Radley, was introduced around 1800 as the seasonal burning of heather strips to encourage new growth. As well as helping to establish a heather monoculture, another by-product of the burning – and the increasing preponderance of wildfire on the moor – was ash deposited on to the upper layer of the peat. For a man interested in the remnants left behind by previous civilisations, this ash stratum was to be the hallmark of modern man, our own layer of history, marking the onset of the Anthropocene.

In some parts of moorland ravaged by wildfire, the heat has baked the soils yellow and red. Elsewhere so much peat had been burned away that it has revealed the barren, periglacial features of the land as it would have appeared at the end of the last Ice Age. These stark moorland tops prove a prescient warning of the extent to which our actions (and their consequences) are warping the natural passage of time. Although one we have failed to take heed of.

Fast-forward to the present day and the new wave of wildfires Jeffrey Radley sought to explain have become an annual occurrence, an individual season all in itself. By April 2019, Britain had already experienced a hundred wildfires that year, eclipsing the previous high of seventy-nine across the whole of 2018. The fires stretched from Cornwall to the Scottish Highlands, and the Peak District was once more ablaze.

★ ★ ★

Towards the end of our time in Pedrógão Grande, Bruno and I visited the village of Ferraria de São João, a few miles from Nodeirinho, which had been spared the worst of the wildfires. A tiny community of forty or so residents living in stone cottages built out of the quartzite ridge it sits upon, Ferraria, as it is known, is one of the most beautiful villages I have ever seen. Goats clattered down the cobbles of the main street, living in animal sheds bedecked in wildflowers while the traditional red clay tile roofs were surrounded by a sea of green from the surrounding forest.

On the drive into the village the eucalyptus monoculture gives way to an ancient cork oak woodland. When wildfires swept down the valley it is these trees that saved the village. Cork oak is a tree native to Portugal whose thick bark makes it especially tough and resistant to flames. In 2017 the cork oak forest worked as a natural firebreak.

In the aftermath villagers acted quickly to ensure their safety before the next wildfire. It was agreed to implement an 'Environmental Protection Zone' around the village in a 500-metre-deep circumference, felling any existing eucalyptus trees and planting oak, cork oak, walnut and cherry to bolster the woodland. Some of those behind the scheme gave Bruno and me a tour of the new orchards and explained how they were harvesting the trees for produce to eat and sell. They have been assisted in their efforts by some five hundred volunteers, including many from abroad, with visitors urged to adopt a cork tree to help protect the village in the future.

The plantations were an exercise in returning to a more natural balance with the land. During the years when they sustained thriving communities, each of the villages in this area would have been similarly surrounded by orchards and olive groves. In restoring what had been, they were breathing new life into villages that had been hollowed out by modernity, and in the process providing protection against the worst ravages of a changing climate.

When I met Kate Hanley on the one-year anniversary of the Saddleworth Moor fire, she was eager to show me similar work taking place to help heal the land. In 2010 the RSPB established a partnership with the water company United Utilities, which owns 4,000 hectares of the moor. Restoration work had already started on the moor prior to that agreement but it remained in a dire state.

Over recent years much of the focus has been on raising the water table and reversing decades of disastrous policy across the northern uplands, whereby the moors were drained to ensure water flowed more rapidly into reservoirs. At the time this was presumed the best way of managing the land but in recent years has been shown to severely impact its resilience and increase the flood and wildfire risk to communities living below, as the destruction of blanket bog means the land is unable to store water. The quality of the water that ends up in the reservoir is also much reduced as it has not been filtered through the natural peat and requires expensive chemical treatment before entering our homes.

Peatlands are vast carbon sinks in their natural state but draining (and burning) turns them into a major source of greenhouse gas emissions, estimated at 1.3 gigatonnes of CO_2 annually, equivalent to 5.6 per cent of global anthropogenic CO_2 emissions. In the Peak District alone, 20 million tonnes of carbon is stored in the peat. The conservation partnership Moors for the Future has conducted an analysis of a 2018 wildfire on a stretch of blanket bog known as the Roaches in the Peak District. Of the 61 hectares burnt, it is estimated nearly 11,500 tonnes of carbon dioxide were released into the atmosphere (the equivalent of running 1,426 homes for an entire year).

Up on Saddleworth, Kate and her team have installed around eight thousand so-called 'gully blocks', anything from hay bales to stone dams inserted into the moorland streams to help raise the water table. On the parts of moorland where this has happened, populations of threatened moorland birds, including dunlin, golden plover and curlew, have all significantly increased, as have species such as crane fly, which require constant moisture to breed. In 2004, she tells me, there were just seven pairs of dunlin in the entire area but by 2018 that had increased to forty-nine pairs.

We meet at Dovestone Reservoir before heading up to the tops together on a narrow moorland path. I tell Kate I have been here before, in early 2015, when the body of a middle-aged man was found near the top of the road with the poison strychnine in his pocket, and police launched a huge investigation to establish his identity.

I wrote an article attempting to piece together what had happened but it was not until two years later that he was discovered to be a 67-year-old man called David Lytton. It later emerged at an inquest into his death that he was originally from London but had moved to Pakistan in 2007 without telling his former girlfriend, who had previously suffered a miscarriage of their baby. He returned to Britain in 2015, took a train up to Saddleworth, where he called in at the last pub before the moors, The Clarence Hotel. During the 1960s, the Moors Murderers Myra Hindley and Ian Brady (who buried their child victims on Saddleworth) supposedly used to stop off here too and would bet sixpence on games of dominoes in the snug. On that night in 2015 the man asked the landlord 'the way to the top of the mountain'. He pointed up the street into the darkness.

The RSPB wardens on the moor were among the last to see David Lytton alive before he was discovered by a cyclist the following day. One of the initial theories pursued – and later discounted – by police was that he may have been a childhood survivor of a plane crash close to where his body was found at a towering 1,500ft (450m) rock formation called Wimberry Stones. On 19 August 1949, a British European Airways Dakota DC-9 plane crashed here in heavy mist, claiming twenty-four lives. Two young boys were among the eight who made it out of the wreckage alive. Eventually the 2017 inquest recorded an open verdict into David Lytton's death – nobody could say what drew him to end his life up here.

On the way up to the moor we pass new saplings Kate and her team have planted to increase the fire resistance and biodiversity of the landscape. On several occasions many of the young trees have been lost to fire before they can properly take hold, she tells me, the frustration rising in her voice; not least as the majority of those fires have been started deliberately. During my research into wildfires I discover to my surprise that this is nearly always the case. As Professor Chris Evans, a senior researcher at the UK's Centre of Ecology and Hydrology, told me: human interventions in the landscape and extreme hot periods make it more conducive to fires happening, but it is almost invariably people that light the spark.

Some start fires out of carelessness, discarded cigarettes or barbecues lit on the moor. Others do so deliberately to clear land, which, in a tinderbox environment, can quickly spill out of control. And then there are the pyromaniacs, drawn by an obsessive impulse to watch the world burn. Even as firefighters struggled to contain the Winter Hill blaze, a police helicopter surveying the area spotted arsonists attempting to light other areas of grassland nearby. The Saddleworth Moor fire was also treated as an arson incident though, despite several arrests, nobody was ever convicted.

One of the emergency responders I spoke with who was on duty at both Saddleworth Moor and Winter Hill was Craig Hope, the lead wildfire officer for South Wales Fire and Rescue Service. Craig is part of a group of twenty-five specialist advisors who provide tactical support when wildfires break out across the country. He told me 96 per cent of wildfires in South Wales are started deliberately and he believes a similar figure is the case across Britain. Often when a wildfire is burning it will attract pyromaniacs from around the country to travel over and keep the flames going. 'We don't know why people do it,' he told me. 'We've tried to analyse the mindset [of arsonists] but we just don't know. Personally, I think it is all kinds of people of different ages and from all walks of life.'

Kate and I have reached the top of the moor and as we get out of her car we hear gunshots ringing out. We realise that by chance we have agreed to meet on 12 August, known as the Glorious Twelfth in the shooting community, marking the moment when the grouse season gets under way. We head in the other direction away from the guns, sloshing over the system of hummocks and hollows Kate and her team have dug into the moorland to improve its capacity to store water. Peat-rich water soaks up over my trainers, which I instantly regret wearing. The newly formed watercourses are surrounded by tussocks of cotton grass, their white buds furled up by August. Cranberry and cloudberry are thriving here too.

As we walk further through the bog the patches of sphagnum moss that have been planted become more apparent. The moss is brought over from Cumbria and Wales then planted in plugs into degraded peat using a tried and tested technique which Kate demonstrates: squashing a hole with the heel of your boot. I drop to the ground to inspect the moss and close up it reminds me of the bed of a coral reef. The mosses are bright stars of brilliant yellow and green, capable of holding eight times their own weight in water.

I dig six inches down into the rich black peat and can make out fragments of old sphagnum which Kate tells me could be up to 400 years old, preserved by the acidity of peat soils, which acts as a sort of time capsule. Kate has discovered centuries-old tree trunks in the bog. Looking out over the largely treeless landscape, which due to the restoration work is now slowly starting to rewild itself with saplings of willow, aspen and rowan, Kate says the discovery is a glimpse of how these deforested hills once might have looked. 'That was humans that cut them down,' she says.

Near to the edge of where the wildfire finally stopped, its progress slowed by the newly wetted moor, we discover purple-stained hare and fox scats, proof that the animals are returning. Kate says what her work here has taught her most of all is the ability of the land to heal itself, if only humans give it the opportunity to do so. 'What you see really shapes your perceptions of what is right,' she tells me. 'Perhaps it is also about letting go of what you think should be there and letting the landscape dictate.'

I wrote that last sentence down in capital letters in my notebook and it stayed with me over the months that followed. That summer of wildfires in 2018 also coincided with the moment where we started to realise that if we were going to have a baby then it was not going to be straightforward. It was a date that marked the point when we had been trying for more than a year and it marked in our minds a shared understanding that something might be amiss. After assuming for the ten years we had been together the rough trajectory our lives might follow, the thought was beginning to gnaw at the edges of our minds about what might not be.

I had always looked forward to being a dad and everything that entailed. As the prospect started to retreat from me it was the details in my own imagination that I found hardest to bear. I found myself reliving the happiest memories of my childhood but this time with me as an adult and nothing in my place. I pictured myself not taking our child to its first Arsenal game, as my dad had done with my brother and me; family adventures we wouldn't have, such as heading out tramping across the moors as I used to do with my grandparents when we dammed up the streams (as the Saddleworth conservationists are doing today). Then there was the thought that neither of us might ever be able to hold new life in our arms.

I can only write here of my own inner feelings of that painful time. Liz and I both spoke in depth to each other about what we were going through and how we were coping. I marvelled then, as I have always done, at her reserves of inner strength and ability to see the good in all manner of situations. On occasion that summer my mind lurched and I relied on her to keep me afloat. I hope I was able to do the same for her.

Watching the landscapes of the Peak District and Portugal renew themselves also provided some solace. I had witnessed nature take its course in the most vicious way imaginable but in the recovery of the moors and mountains I had also appreciated its durability. In the sturdy cork oak trees whose bark resisted the flames and the constellations of sphagnum moss newly spreading across Saddleworth I saw the ability of nature to flourish where hope has seemingly been lost.

I held on to this as a sign that whatever was going wrong between us might one day, one month, be suddenly corrected and we would manage to conceive. But after visiting Saddleworth with new life slowly emerging all around me, I also considered for the first time that perhaps our story would not ultimately be about the child we were – or were not – able to bring into this world, but instead the way we ourselves responded to whatever nature had dictated for us. For love, like life, is nothing if not resilient. Liz taught me that.

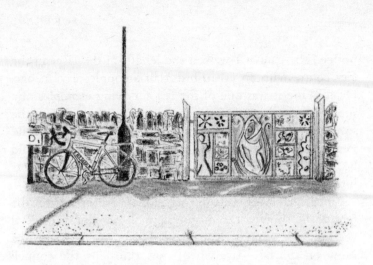

CHAPTER TEN

Melting

A mile or so from my home is a relic of winter. It stands on the edge of the Sheffield ring road, facing out across the traffic that runs past the main railway station. Alongside the thundering cars there is a glimpse of an open stretch of the River Sheaf – the river from which Sheffield takes its name – which otherwise has been largely built over in the city centre, encased in concrete culverts. The river comes up gasping for air for only 50 metres or so before it plunges into a dark brick tunnel which leads downstream to feed into the River Don. The Pulp singer and son of Sheffield Jarvis Cocker wrote a song about floating in a raft down this underground stretch in his youth.

Where the water is exposed to daylight a wooden gate is built into the brickwork along the river bank. The slats are covered in graffiti and metal fixings long rusted. I first noticed the gate as I walked the length of the river in the city centre with an urban historian a couple of years ago. He told me it was one of the last surviving examples of a snow gate in the city. The gates were necessary due to the severe winters that often swept in from across the wild Pennines. In times of heavy snow, the wooden slats would be lifted out to enable the workmen to shovel the drifts into the river.

Since that visit I have occasionally returned to the redundant snow gate, imagining with an almost visceral yearning the winters that once held my city in their grip. Now on the rare days when snow threatens, the council dispatches its gritting fleet instead. Even in mild winters so much gets spread that by spring my bicycle chain has rusted a slug orange.

Most of Sheffield's snow gates, like its snowy winters, have disappeared: stoved in by vandals or obliterated in asphalt in the name of road improvements. Further upstream the Sheaf, on Saxon Road, three metal memorial gates have been designed by a group of artists to commemorate the snow gates that once stood there. A poem called 'Equilibrium' inscribed on one considers how human activity interrupts the natural passage of the seasons: 'Balance is the key to life struggling to keep sane/ To tear it with a knife would make the balance turn to rain.'

Much is rightly made of the loss of Britain's biodiversity in the modern era, but what of the loss of our weather diversity? While we have more extreme events, in the manner in which some rare migrant bird might blow in for a few weeks sending twitchers scurrying after it, overall our weather is becoming duller, greyer, warmer and wetter, and with far less of the pronounced scene changes one typically associates with the four seasons. Last winter my neighbour, who festoons her home each year in kitsch Christmas lights, made an addition to the festive array: a projector that beamed snowflakes on to the facade. Usually I love the lights, but I found this artificial snowstorm during a winter in which there was none uniquely depressing.

Alongside our changing climate, human ingenuity has worked to de-weather our cities; to nullify the elements as best we can lest they interrupt our daily charge. In the opening pages of *Bleak House*, my favourite weather writer of all, Charles Dickens, describes a November day in London where the soot falls like snowflakes, mud splashes horses up to their blinkers, and entrails of fog creep along every side street and down every chimney pot. The weather interjects like great paint splodges on the canvas of city life. So where is it today?

In December 1813, such a fog as Dickens describes swallowed London whole. Foreign Secretary Viscount Castlereagh, whose ministerial carriage was hastening to Germany where allied forces had massed on Napoleon's borders, was forced to travel from 18 St James's Square to Harwich, Essex, at a snail's pace, navigated only by the torches that dangled over the heads of his horses. Hackney coachmen careered out on to pavements, drawn like moths to the dim lanterns. The Birmingham mail coach took seven hours to travel just 20 miles from central London to Uxbridge. On 28 December, the returning Maidenhead coach missed the road near Hartford Bridge and overturned. This 'tremendous fog', as one contemporary chronicler recorded, grew so dense that by New Year's Eve the gas lamps flickered no brighter than candle stubs in cracked basement windows.

Then, carried along by bitter north-easterly winds, came the snow, falling for forty-eight hours straight. As temperatures plummeted, 2-metre-long icicles took perilous form on buildings and London came to a standstill. History neglects to tell us much of the poor who presumably froze to death in squalid tenements, unable to work and with no money for food, but there are florid accounts of the rich donning fox furs and skating across the Serpentine.

The weeks passed and frosts hardened until the River Thames froze solid. The Thames ferrymen, those ragged forms immortalised by the nineteenth-century painter Whistler, quickly spied a business opportunity and hauled a sheep on to the ice to roast over a fire. To stand and watch the 'Lapland Mutton' cost sixpence, while the meat was sold at a shilling a slice. The last, and perhaps greatest, of the famous Thames Frost Fairs was born.

Within a matter of hours, booths decorated with streamers and flags popped up to sell beer, gingerbread, gin and meat. By the start of February all life poured on to the ice to drink, dance and gamble. At one stage, even an elephant was led across near Blackfriars Bridge. The fair lasted a week. By then the cries of revelry had become interspersed with ominous cracks across the surface. Those who had set up booths were desperate to extract every last penny they could and remained open in defiance of the warnings that the ice was breaking up. During these final throes a number of printing presses and fair booths were swept away downstream. So too a pair of 'genteel-looking men', recorded drowned off Westminster Bridge.

When the last stragglers finally hauled themselves on to dry land it would prove to be the end of London's frost fairs. The medieval London Bridge, which was demolished in 1831, had boasted nineteen narrow arches that slowed the flow of the Thames. Its replacement, plus the creation of the Embankment, created a narrower, faster-flowing river. More importantly, towards the end of the nineteenth century, Europe began to warm.

★ ★ ★

The Central England Temperature (CET) series is the longest meteorological dataset in the world, spanning 350 years. Administered by the Met Office, it records daily and monthly temperature readings representative of a roughly triangular area between Lancashire, London and Bristol. Its monthly records date back to 1659. The CET has also been recording daily minimum temperatures since 1878. We may still culturally hark back to the Victorian winters Dickens described as the archetype of what the weather should be like today, but the CET reveals that in fact they are more like a distant ice age. Here is concrete scientific evidence of that implacable feeling I have each winter as I wake up to another grey morning and wonder where have all the frosts gone?

Searching through old weather columns I have written over the past decade, I find numerous reports of these changing winters: 2014 witnessed the lowest number of spring frosts ever with a mere twenty-three days recorded. In October 2015 I wrote about the premature arrival of the Bewick's swans at the Wildfowl and Wetlands Centre at Slimbridge, Gloucestershire. The birds migrate from the Arctic each year and in Britain have long been associated with the beginning of winter. There is a saying that they 'bring snow on their bills' because flocks tend to move just ahead of biting cold snaps.

The Bewick's swans are not the only bird steeped in folklore for helping predict the severity of forthcoming frosts. 'If ducks do slide at Martinmas,' so an old saying goes for what the weather is doing on 11 November, 'at Christmas they will swim. If ducks do swim at Martinmas, at Christmas they will slide.' When the first Bewick's swans touched down at Slimbridge that October in 2015 it was the earliest date since records started in 1963 and a full twenty-five days earlier than the previous year. The swans, I wrote, prompted hopes that a proper winter could be in store. In fact, the winter of 2015/2016 transpired to be the warmest for 147 years.

One December afternoon sultry enough to make the Bewicks consider heading back to the Arctic early, I arranged to speak with Dr Mark McCarthy, science manager of the Met Office's National Climate Information Centre, and a man responsible for overseeing the CET. According to Mark, the data suggests a significant decline in frosts – around 15 per cent of both air and ground frost records compared to the baseline climatology period of 1961 to 1990. This warming process has significantly increased since the 1980s. The list of the coldest twenty months in the CET series to 2011 features only two other months since 1988 and they were summer anomalies, May 1996 (the fifteenth coldest May on record) and June 1991 (the eighth coldest June on record).

The twenty-first century so far, Mark told me, has been warmer than any equivalent period in 350 years of instrumental records. Against this backdrop of rapidly rising temperatures, we will still experience occasional winter extremes, such as the exceptionally snowy 2010 (forever stamped in my mind as I wore thermals to a nightclub in Camden and ended up sweating so profusely I had to undress in a toilet). But overall we are living through what is considered to be the end of winter. According to the latest Met Office research, if global greenhouse gas emissions continue to rise at the present rate, by 2040 most of southern England may never experience days where the temperature remains below freezing. And by the end of 2080 only areas of very high ground and some parts of northern Scotland may experience freezing days. Come the year 2100, snow settling on the ground will be a thing of the past.

As we talk, Mark describes what he calls the 'chronic impact' of our warming climate. This is more about the cumulative effect of milder winters and a reduction of frosts on the surrounding landscape, nature and us. How that changing weather will manifest itself on an annual basis and what might the possible impact be? An example Mark cites is that perhaps one day soon we will be mowing our lawns throughout the winter. A lawnmower droning through suburbia in December may sound trivial, he admits, but the potential consequences of this warming are immense.

A few weeks after we speak, Cassley in Sutherland recorded 16.8°C at 3am on 29 December, the highest UK temperature ever reached so late in the year. I went for an evening walk in the woods near my house where tawny owls were hooting under the light of a waning moon and it was so warm I didn't even need to do up my coat.

That was the final record broken for 2019, a year described by forecasters as 'exceptional' for the number of extremes registered. That word in itself fails to do justice to what is happening: the true exception would in fact be a winter of proper prolonged cold rather than what seems to me a depressingly typical example of the future that lies ahead. So many records are these days being broken that perhaps it is time to rewrite the record books, and accept the aberration has become the norm.

On 25 July of that year, Britain witnessed its record high when the temperature at Cambridge Botanic Garden reached 38.7°C, surpassing the 38.5°C registered in Kent in 2003. On 26 February we experienced the warmest ever winter day when temperatures of 21.2°C were recorded at London's Kew Gardens.

I visited Kew on the balmy morning after that record-breaking February day with a plume of warm southerly air still sweeping up from the Canary Islands. I took a stroll through the gardens with Richard Barley, director of Horticulture, Learning and Operations, who had agreed to show me some of the effects of that record-breaking spring when nature was going haywire.

To illustrate the impact of the unseasonal warmth he took me to the shadow of the Romanesque campanile tower by the main entrance, which disguises the vents from the boilers that heat the Palm House. The hot water pipes run underneath the earth between the two buildings, and Richard pointed out to me the allium bulbs closest to the pipes, which had already grown an inch taller due to the warmth. Every year the bulbs nearest to the pipes come into bloom first, but that February everything had exploded prematurely into life. Next to the plant house a mound of daffodils was already in full bloom. Richard told me the Kew data records show that since the 1980s the daffodils have been coming into full flower several weeks earlier as longer springs subsume ever more of winter.

Of course, such early blooms signify more than the demands of our own cultural expectations. Many of Britain's plants have evolved to use winter's cold to their advantage, requiring a period of dormancy during the colder months – called vernalisation – to reset their biological clocks in preparation for the spring. The energy required for plants to come into bloom earlier and for longer can reduce overall lifespan and cause chaos among pollinators.

We walked past a copse of magnolia trees whose pale precious flowers were just beginning to emerge from the hairy buds in which they had been sealed the past few months. Richard admitted over the sound of puttering sprinklers on the main lawn and squawking parakeets that he had never known a February like it. A few days previously at Kew he had seen a man sunbathing on the grass with his shirt off. And to think that same month two centuries ago, a couple of riverbends downstream, Londoners hauled an elephant across the frozen Thames.

★ ★ ★

Sheffield may have the ruins of its snow gates, but in other parts of the country whole industries have crumbled as what the poet John Clare described as the 'hoary shroud' of winter has lifted, perhaps for good. What that change reveals is a season that lies in memory rather than experience. Take the curiously shaped gritstone boulders near to me in the Peak District which, according to local legend, were once the bodies of agricultural labourers frozen solid in the frosts as they toiled. Such stories stand as monuments to a country and culture that has been shaped by the weather. And now the great thaw is once more reshaping our communities and lives.

Centuries ago the extent to which we could rely upon winter being long and harsh was evident in the number of frost quarries that became a viable industry in parts of the country. The aptly named Cotswolds village of Stonesfield was one such example. In the sixteenth century local quarrymen discovered that the belt of Jurassic limestone underneath the village was particularly susceptible to splitting following winter frosts, which made it perfect for sculpting into roofing slates.

At its nineteenth-century peak there were some two dozen quarries mining the stone and the work was entrenched in the annual calendar of the village. 'They looked forward to it almost like the annual migration into the mountains by the Alpine shepherds,' the Oxford geologist and palaeontologist W.J. Arkell wrote of this subterranean transhumance.

The quarrying began at Michaelmas and continued until Christmas with blocks of stone hauled up the mineshaft to the surface, usually 6 metres or so above. Then they were left to weather. On cold nights local men of the village were employed to wet the stone slabs to ensure they froze. One week of hard frost in January would provide the 'slatters' (the men who shaped the roofing tiles) with enough work to keep them busy until the following Michaelmas. So vital was the weather to this process that when a hard frost was forecast, church bells would ring out to summon workers from their beds.

There was good money to be made. As well as their regular salaries the quarrymen sold the fossils they uncovered to private collectors. By the time the last frost quarry closed in 1911, Stonesfield slate adorned the roofs of numerous country houses and several Oxford University colleges.

In the village of Collyweston, eighty or so miles away in the Northamptonshire countryside, a similar method of using the frosts to split local quarried limestone had built up over the centuries and by some estimates stretches back to Roman times. During the nineteenth-century boom, quarries were dotted in parcels of land leading off from the main road, the biggest of which was on land owned by the Burghley estates and known as the Deeps. As the frosts subsided, the industry also became considered no longer viable and the final quarry closed in the 1960s.

Collyweston stone was used to slate the roofs of some of Britain's most prestigious buildings, including the Guildhall in London, Apethorpe Palace, several Cambridge University colleges, and a nineteenth-century manor house on Long Island, New York. Even after the closure of the mines, the stone remained in high demand although supplies were running critically low.

Recently the scion of a Collyweston slating dynasty decided to do what had previously been regarded unthinkable: reopen a 600-year-old mine and begin once more quarrying for slate. The difference this time, with frosts no longer able to be relied upon, was the quarried stone was to be split inside industrial freezers instead.

I wanted to see the rebirth of this industry and make sense of the idea that having lost the weather upon which we once relied we could now somehow devise artificial means to replace it. Like a snow cannon blasting on to a green alpine piste or Qatar's air-conditioned football stadia conjuring artificial breezes in the broiling desert heat, I wanted to visit an English village that was recreating its frosts.

Typically, of course, the weather had its own say and the November morning I arrange to meet Nigel Smith at his Collyweston quarry coincided with the hardest frost of the year. It had been -3°C when I left Sheffield that morning and as I walk across the builder's yard to his office in a low-rise building next to the quarry, iced puddles creak satisfyingly underfoot and my breath blooms like a speech bubble in front of me.

I am quietly thrilled. Here, after what had been a depressingly wet and mild month, is the 'lusty winter' Shakespeare once wrote of. When Nigel greets me at the door I can tell he is similarly excited about a proper frost. I suppose it is a feeling that has been passed down through generations of Collyweston quarrymen. The cold lies deep in his bones.

Still, this short frost snap was not enough to impact upon the stones, which require far more prolonged exposure. He tells me that when the quarry reopened he had hauled out a section of Collyweston stone on to the yard to treat it in the old way to see if it might still split. He had checked that morning. The stone had not fissured even a feather's edge.

Over tea inside with Nigel and his wife, Viv, the pair explain to me their shared love of Collyweston stone. Nigel's great-grandfather was a slater (though after injuring himself falling off a roof later ran a pub in the village) and his grandfather and father, too. He grew up under Collyweston stone, in a house on the main street of the village whose roof was adorned with the slates, and entered the family business at the age of sixteen. By then the quarries had closed and the work was in repairing and re-purposing Collyweston slates salvaged from derelict farm buildings in the surrounding countryside.

While working up on rooftops in the early 1990s, Nigel says he first started to notice how much warmer the winters were. The fields below him, typically so susceptible to hard frosts from weather blowing across the Lincolnshire plateau, were increasingly remaining free of ice for extended periods of time.

Nigel was immersed in the quarrymen's stories in his youth and delightedly recounts them now. There were once four pubs in the village (his great-grandfather's among them) and on frosty nights the workers would take it in turns every hour or so to carry pails of water up to the fields to pour over the stones. On a wall in his office he points out to me a photograph of his great-grandfather taken in the early 1900s when he was commissioned to take a shipment out to New York for the Long Island manor house. Nigel's firm is currently re-slating what was known in the US as the Golden Roof, and one of his team of slaters working out there is a man called Tom Measures, the great-great-grandson of one of the original slaters.

Nigel's father, Claude, bought the land upon which the derelict quarry stood in the 1990s. The only way into the quarry back then was through a narrow mineshaft. With the industry dead, Claude simply used the land as a builder's yard. The rich seam of Collyweston stone beneath their feet was left to a colony of pipistrelle bats.

But in 2014 Nigel heard of an experiment arranged by Sheffield Hallam University in conjunction with Historic England to devise new means of creating Collyweston slate due to a chronic national shortage. The scientists discovered that artificial processes could recreate the freeze and thaw techniques once provided by the weather. The following

year Nigel applied for planning permission, despite what
Viv wincingly describes to me as the astronomical sums
involved. As soon as permission was granted he began work
on reopening the quarry.

We walk out to the mine workings fringed by a radius of
buddleia spanning 20 metres or so. Nigel says when the new
entrance was excavated these pioneer plants quickly reseeded
themselves in the ploughed earth and in summer transform
the mine into a joyous ring road of butterflies. Beside the
mine are four industrial freezers (Nigel secured the first one
for his burgeoning business second-hand from the back of a
Tesco lorry) where the quarried stone – known locally as
'log' – is treated. Generally it takes three cycles of frost and
thaws to split a log. In between freezing, the logs are kept in
large plastic vats, soaking in water pumped directly up from
the cave as tap water reacts with the limestone to create
algae blooms. Even though the freezing now takes place
artificially, Nigel tells me, the stone still splits far better in
the winter months, although no scientist has been able to
explain to him why this should be.

The entrance to the mine looms below us and as we
descend into the gloom the air stills and cools. Often in the
morning when Nigel first opens the door to the mine, an
eerie fog rolls out and over his feet. Tiny stalactites have
formed above our heads. I touch one and it dissolves into
silty residue on my fingertip.

The Collyweston seam is sandwiched between a ceiling
of limestone bedrock 3 metres thick and soft sandy clay
below. The slate is veined with blue streaks that slowly turn
a gun-metal grey as it oxidises. Standing underground
gazing upwards I think of the centuries that man has mined
here and how for all that this narrow seam of rock has
shaped life in Collyweston above the earth, in the billion-
year history of geologic time, that same period represents
merely a flicker.

As we head deeper inside a bat flutters in the gloom between the old quarry workings from previous centuries – distinguishable by piles of stone propping up the ceiling rather than the reinforced pillars of the modern mine. Men like Nigel's great-grandfather would be down here on their backs scratching away with a hammer and chisel in the candlelight for hours each day. The work was perilous, and being buried alive under a collapsed section of stone was a constant risk. 'We don't know of anybody who died, but you don't hear those stories,' he says. 'I'm sure there must have been some down here that did.'

Now the log is loosened remotely by a machine called a Brokk 100, which hammers into the densely compacted sand underlining the slate. The Brokk has turbocharged the process, helping excavate about 200 tonnes a year. When Nigel opened the mine he was told he had about ten years of quarrying before the seam ran out, and estimates he is about halfway through. Where the slate is exhausted the mine is left to the bats, with special roosting cavities constructed into the earth and insulated soundboards installed to deaden the sound of the Brokk 100 hammering away at the existing slate.

We leave the mine to its bats, and ghosts, and head back up into the daylight to inspect the stone yard. Blocks of log are piled up waiting to be split, covered that morning in a thick rime which highlights the grey shades of the slate like stain on wood. I can hear loud banging from a cavernous shed at the end of the yard where the slate is being cut. They call it cliving, Nigel points out to me, in the same patient manner as when I erroneously call the slates, tiles. There is a whole vocabulary around the Collyweston frost quarries which he has rescued alongside the tradition and is fiercely protective of it. The weather here had created a new and unique language which was at risk of dying out with the frosts.

At the time of my visit, Nigel has twenty-three people employed on site. The cliving shed is a hive of activity, with a radio blaring out the Cranberries and men in high-vis jackets pounding away under electric heaters. I am taken along the production line and shown how the slate is split down the seam and cut into three square sides with a heavy-duty electrical saw.

A couple of workers chip away with dressing hammers to give each slate an authentically distressed look. Then a nail hole is drilled into each slate to enable it to be secured on the roof – a process once done using something called a bill and helve.

There are twenty-eight different sizes of Collyweston tile, each with its own name and counted in the old measures. The largest slates, used in the eaves, can be up to 24 inches wide by 36 inches long. Along the edge of the barn, away from the hammering and sawing, the finished slates are kept in containers destined for the roofs of various Cambridge University colleges, where Nigel's team is currently working. 'People said Collyweston was finished, done,' he says to me, beaming with pride at the surrounding din. 'But I've done something here. I've proved this is possible.'

My tour over and feet numb with cold, we part ways. As a farewell gift Nigel gives me two Collyweston slates. One square with a stain across it where the stone is yet to oxidise; the other with a sharp jagged ridge at the top that glitters where the light catches it. I have nailed them both above my bath and sometimes lie there in the hot water thinking of the way the frost seeps deep into the slate. Stone and skin: it serves to remind me how the weather shapes us in curious ways.

After I leave Collyweston I reflect on all Nigel has saved through reopening his quarry. Not just the fact that newly mined slate will adorn roofs for another century, but the maintenance of everything that built up alongside the industry. The unique skills required to shape the stone, the vocabulary of the industry utilised by a new generation, and how re-establishing a connection to the past has excavated centuries of stories and history alongside the stones.

And yet intrigued as I was to see the Collyweston slate revival, this was tempered by a sense that the basis for this rebirth was artificial. While the stone in Nigel's freezers split far quicker than ever before, I thought also of the slabs he had left out in the cold to see if the natural approach would still work and the fact they had remained unmoved.

Frosts formed the basis of this industry and now they have become a mechanised part of the process. We can recreate one aspect of winter proper but what of the wider impact of losing our frosts? What of the trees no longer able to rely on a dormant stage or animals unsure when to hibernate? What of the impact of the loss of the frosts on our own bodies? What natural process might be being disrupted within ourselves?

★ ★ ★

These thoughts came into sharper focus a few weeks later on New Year's Eve sitting under the strip lights of room seven in our local doctor's surgery. Another batch of test results had come back and once more the doctor told us they were inconclusive. We were going to be referred to the gynaecology clinic of the nearest hospital for further examination and to discuss what our options were.

Our GP was young, probably mid-thirties, a similar age to us, and as she spoke fixed us both with kind eyes that made us think she had perhaps endured something similar in her own life but was too professional to say. Try taking evening primrose oil tablets, she told Liz as an off-the-record aside;

they taste like garbage but apparently sometimes help people conceive.

There were other options we were also now being forced to consider, about which we were both deeply unsure. We had already discussed IVF and both had reservations, not just the obvious intrusiveness of the procedure but also the pain of renewed hope.

That is, of course, not to say IVF isn't the right choice for some in our situation and I have friends for whom it has happily worked first time. But we felt we could be drawn down a path from which, once we had set off, we might not easily be able to return. As a journalist, one of the most painful interviews I have ever done was with a lovely couple who had endured seven failed rounds of IVF in pursuit of a pregnancy which always remained just out of reach. They were tough, determined souls and looking to adopt a child instead, but as we spoke the experience was etched on their faces.

Our doctor was kind enough to fill in and send off the referral form right in front of us as an assurance that we were being treated as a priority and wouldn't fall into some NHS rabbit hole between departments. She read out loud as she entered our most intimate information into the grey boxes on the form and the word 'infertility' stung my ears – I suppose, like any medical diagnosis, when you hear it in relation to yourself it takes on new meaning.

Back at home I wandered around our garden to gather my thoughts. It had been overly heated in the surgery and I wanted to be out in the late December weather to restore some sense of balance to my own mind. But even here outside it was stuffy and warm, as it had been for days. Inspecting the plants to distract myself I noticed the first catkins had flowered on the two hazel trees in our garden, hanging like lamb's tails from the ends of the twigs, when typically they should wait for another month at least.

It was early afternoon but already the light was fading on the end of another year, our eleventh warmest on record, and the end of a decade which broke eight British records for high temperature extremes. That night we celebrated at a party with friends. Anxiety was easy to disguise among the distractions of New Year revelry, but in truth I was afraid about what lay ahead.

CHAPTER ELEVEN
Waterland

Christmas is often a fraught occasion for a journalist and that of 2013 was no exception. I was on call over Boxing Day, watching the rain hammer down outside the window of my flat first thing in the morning, when my phone rang. An editor on the other end of the line offered brisk festive greetings and the breaking news that the three rivers that converge at the Kent village of Yalding had burst their banks. I was to get there as soon as possible.

Yalding, a pretty village with a teapot museum, some fine examples of oast houses and the longest medieval bridge in the county, had previously flooded in October 2000. Dozens of homes were evacuated and the repair bill stretched to £6 million. Villagers were told that particular deluge was a so-called one-in-a-hundred-years event, meaning there is a 1 per cent chance of it being equalled or exceeded in any given year. In the end it only took thirteen years to top it – and the ensuing damage was far worse.

In the early hours of Christmas Eve 2013, following days of heavy rain, the rivers Beult, Tiese and Medway breached, sending a muddy torrent surging down the high street. Soon the entire village was plunged into darkness as electricity cut out. By Christmas Day, residents were left calling out for help from upper-floor windows to rescuers in boats. In total, around a hundred houses were evacuated, as well as the nearby Little Venice caravan park, with people taken to the church and village hall. The postmaster had to be rescued from a window as his home flooded. I met him standing outside inspecting his Royal Mail letter sacks stained dark red by the damp and writing a sign that read: 'Closed until further notice'. Flotsam of the festive season was everywhere I looked.

After wading into a quiet cul-de-sac, I was shown around the ground-floor flat of an 82-year-old who had just been airlifted to nearby Pembury Hospital. He had been rescued by a group of neighbours on Christmas Day, who had forced their way through the back door wearing fishing waders. They discovered him sitting in his armchair up to his waist in the filthy, freezing water. As I walked through the ruined house the carpets oozed an evil black silt and paint had blistered along the skirting boards. In the kitchen, the force of the surge had hurled his fridge-freezer on to its side.

Elsewhere in the village I met Joseph Wilson, a then 17-year-old Sea Scout, who had rescued ten different trapped residents in his kayak. Another resident showed me four chickens she had managed to retrieve from their coop in her back garden and was keeping safe in the bath. The owners of the teapot museum, meanwhile, had retreated to their home above the shop where they cooked a Christmas dinner of gammon steaks on a skillet over a wood-burning stove. They later told me that the family dog, Ebony, a black

Alsatian, died of internal bleeding a short while afterwards, which they attributed to the stress.

When the then Prime Minister David Cameron emerged to inspect the damage, he was heckled from the sidelines by residents who wanted to know where he had been for the past few days. One mother-of-three accosted him to say her family had been abandoned in their cottage since Christmas Eve. Cameron muttered something back about ensuring she got all her carpets and furniture out which understandably failed to appease her.

I remember the Prime Minister's words well as he surveyed the raging River Beult. 'You only have to watch the news over the last few years to know these events are happening more often now,' he told the assembled press pack. At the time we had a climate-change sceptic for an Environment Secretary, so even this mealy-mouthed appendage to the floods felt like something approaching a watershed moment. But another question remained unanswered, hanging over the ruined high street: what should people do about it?

The drive out of the village was eerily quiet. I headed towards the main road to Tonbridge, where an old friend lived and had kindly agreed to allow me to write my article from his kitchen. It was beginning to get dark and the road was deserted. Plastic rags deposited by the floodwaters fluttered like ribbons in the hedgerows. I swerved around trees that had toppled down on to the road and a stranded ambulance. The scene resembled a disaster movie, which, locally, is exactly what it was.

When I reached the A26 I stopped at a junction waiting to turn left and suddenly felt an almighty smash and the breath sucked out of me. Winded and in shock I got out and watched the other driver, who had gone straight into the back of me, stagger out of his vehicle, pale and apologising profusely. A quietly spoken middle-aged man, he told me he had been flooded out of his house on Christmas Eve and hadn't slept since. That afternoon he had finally made it out of Yalding to go and stay with a relative, and was so exhausted he had fallen asleep at the wheel.

The back of my car was crushed. The garage repair man later told me if the bill had extended to another £50 it would have been a write-off. Still, once we had exchanged insurance details the engine wheezily burst into life and I clattered my way to Tonbridge.

I was relatively new at the *Telegraph* then and didn't even bother telling my editors I had just been in a car accident because I was more concerned at the 1,000-word-shaped hole on page five of the next day's paper with my name at the top and only an hour to fill it. I remember my friend watching me bash out my story at his kitchen table while fielding angry calls from the newsdesk, surrounded by Christmas decorations and the happy detritus of his young family. He told me he would prefer almost any job he could think of to mine. The next day I woke up to discover my neck had frozen solid with whiplash.

During that visit to Yalding I encountered a strange phenomenon I have witnessed many times since: environmental trauma. A sense of utter helplessness was a shared experience among the victims of the floods; from the stricken village postmaster to the man who fell asleep at the wheel and careered into the back of me. People who had lived in Yalding their whole lives were suddenly being confronted with the threat of an altogether more destructive type of weather that threatened their homes and livelihoods.

In later years, reporting on repeated flooding elsewhere in the country, I would have the impact described to me by specialist mental health teams working with families who had lost everything as similar to Post-Traumatic Stress Disorder. Others liken the feeling of being flooded out to a bereavement. Neighbours and relatives rally round and help each other through the immediate shock but there comes a moment when you are left alone in the ruins of what was formerly your home, facing the prospect of having to start all over again.

★ ★ ★

Humans have long been braced to be swept away. On the ceiling of the Sistine Chapel, Michelangelo painted the sinners desperately fleeing the great flood sent down to wash away the wrongdoings of mankind. Turner's *The Deluge* conjures a similar scene, echoing the long-standing fear of the weather as a form of divine retribution. When exhibited at the Royal Academy in 1813, Turner's painting was accompanied by a few lines of poetry from John Milton's *Paradise Lost*: 'The thicken'd sky/Like a dark cieling [*sic*] stood, down rush'd the rain/Impetuous, and continual, till the earth/No more was seen'. Commenting on the painting, one of Turner's contemporary critics noted 'that severity of manner which was demanded by the awfulness of the subject.'

Long before Kevin Costner's 1995 action blockbuster *Waterworld* (a childhood favourite of mine in which rising seas have flooded every continent and the last of humanity sails the earth), J.G. Ballard published his dystopian 1960s novel *The Drowned World*, in which the author imagined an abandoned London submerged. Increasingly now communities are experiencing what it feels to be inundated. Every year fresh rainfall records are being broken, brutally exposing the manner in which we have dismantled the land's ability to cope with the run-off. Put all these events together and it should sound the alarm for a national emergency; but of course it doesn't work like that. Even now with every winter (and increasingly summer) bringing a fresh wave of floods, we still somehow manage to tell ourselves it won't happen to us.

I've reported on these disasters all over the country. At the *Halifax Evening Courier* I was occasionally dispatched on my bicycle to wherever the River Calder or one of its tributaries had burst its banks, as cars were unable to get through. In my then hometown of Sowerby Bridge, a gaggle of geese was renowned for its daily procession from the river up across the main road, holding up the traffic as it went. One particularly wet year I remember seeing the geese swimming along the high street.

Later, as a reporter on a national newspaper, my journeys into floodscapes took me further afield. After my Christmas in Yalding in 2013, the following winter I stood on the top of Burrow Mump, a 79ft (24m) Triassic-age hill that looms high over the Somerset Levels, looking out over 25 square miles of swamped landscape. Bruised grey juggernaut clouds cruised above me and down on ground level I watched a swan glide gracefully along what should have been sheep pasture. That was the year the aforementioned Environment Secretary got booted out of the job after arriving in a pair of lace-up brogues rather than wellies to survey what was, at the time, Britain's biggest ever pumping operation in an attempt to disperse the floodwater.

I know now from experience that certain things are guaranteed to happen every time there is a major flood. Politicians will belatedly arrive wearing incongruous footwear and concerned expressions on their faces. One of the earliest viral stories the then nascent website Buzzfeed published in 2014 was entitled '21 Pictures of Politicians in Wellies Staring at Floods'. Another inevitability is that the Environment Agency, which is responsible for flood management and prevention in England and Wales, will be slated for its response. In December 2015 – a miserable month which achieved the record of the warmest and wettest December ever in Britain – almost twice as much rain fell as normal, flooding 16,000 homes in northern England. The then Environment Agency chairman Sir Philip Dilley released a Boxing Day statement saying he was 'at home with his family', neglecting to mention that home was actually in Barbados. He resigned the following month.

Another certainty is people will, rightly or wrongly, insist that a lack of dredging of the rivers was responsible. On the Somerset Levels I spotted a typical banner emblazoned with the message 'Stop the flooding – dredge the river!' on a bridge spanning the River Parrett. Scooping out silt and other detritus from the river enabling faster flow can make a difference in places but experts say it is not the solution flooded communities often believe it to be.

Also, people will rally together magnificently. Driven out of their homes, neighbours quickly gather in streets, conduct rescue operations, provide shelter for the vulnerable, and collect and distribute resources. Flood victims commonly describe a curious sense of loss once the waters have subsided, normality resumes and people retreat back into their homes.

Another certainty is rumours will abound of looters targeting abandoned homes and other uninvited guests putting communities at risk. On the Somerset Levels mass hysteria was whipped up over hordes of rats supposedly invading homes after being washed out of their nests. In the summer of 2019, when the Lincolnshire town of Wainfleet was flooded by what was again record rainfall, another animal appeared in the headlines. Oddly it was not two months' worth of rain falling in two days that was blamed on flooding nearly six hundred homes around Wainfleet, but badgers building their setts on the banks of the River Steeping and supposedly undermining the structural integrity of the town's defences.

After reading in the newspapers about the blame being placed on the badgers, I wanted to see for myself how much of the story was genuine and how much was journalistic mischief-making, with reporters attempting to extract a news line to spice up what would be deemed by some editors to be yet another flooding story. I drove to Lincolnshire the week after Whit Sunday, a date in English folklore typically dedicated to early summer revelry. The Whitsun holiday, always the day after Pentecost (the seventh Sunday after Easter), has been observed for centuries as heralding the start of summer proper. It is a time for Morris dancing, cheese rolling, tossing the sheaf, village parades and Whit walks, a northern festivity described by Charlotte Brontë in 1849 as 'joyous' and 'a scene to do good'.

In Larkin's 'The Whitsun Weddings', the poet describes watching snatched glimpses of a dozen betrothals from the window of a London-bound train in languid sunshine that cast long shadows off the poplars. Larkin's poem ends, however, in an 'arrow-shower' of rain and as I crossed the toll bridge from Yorkshire into Lincolnshire, drops began to rat-a-tat-tat on my car windscreen.

I parked outside Wainfleet and walked past the road-closure signs and long snaking rows of firefighter pumps. Wooden fences were buckled under the weight of the water. In one garden, gnomes were floating on their backs gazing up at the lowering sky. Another redbrick detached house at the end of a neatly paved driveway had three model boats in the living-room window, which were reflected on the surface of the perfectly still lake surrounding the property. The curious symmetry of the scene reminded me of a postcard of an Alfred Wallis painting, *Brigantine Sailing Past Green Fields*, which I have stuck to my fridge.

Down at the town's Coronation Hall, which had been transformed into a 24-hour shelter for residents, I spoke to a few people about what had taken place. Gail and Richard, a married couple in their sixties, had been evacuated two days previously. Their home had been in the family for sixty years, they told me, and never previously flooded. At first, hoping they would be spared the worst, they boxed up possessions and reversed their car on to breezeblocks in the garage. But when the electrics shut off they decided it was time to flee. Eventually the pair were rescued by a dinghy. I asked Gail why she thought the town had flooded so badly and, like everybody else I spoke to, she mentioned the badgers, plus the lack of dredging in the River Steeping. I asked whether she considered climate change to be a cause but she insisted she had barely given that a thought. 'Well I never thought about it affecting us,' she said. 'I was just in my own little world.'

The Environment Agency insisted in numerous statements to residents that this was not the work of badgers, nor indeed insufficient dredging, but due to the river banks being put under extraordinary pressure because of the extreme amount of rainfall. But that seemed an explanation few wished to accept. On the street outside the Coronation Hall I spoke with Katherine Groves and her daughter, Rosie, walking in wellies after visiting their nearby house for the first time since the flood. Katherine's father only passed away nine weeks previously, she told me, and had he lived to see his home destroyed it would have broken him. 'It's memories,' she said, tears springing into her tired eyes. 'I know I'm going to have to throw away lots of things which my family has gathered over our lifetime and mean a lot to us. Now they are just gone. The only way they are recorded is in your memory and it means a complete new start of life.'

That heartfelt summary of the misery of flooding perhaps also explains why people affected seem more willing to latch on to simple causes, which we have either created or can fix. The alternative to badgers or a build-up of silt,

I suppose, is an acceptance that our individual and societal actions have helped change our weather, and that in turn is going to change our lives.

★ ★ ★

Then one November evening that same year in 2019, the waters came for me. It had been the worst of months: rain all day, every day, the endless pattering applause of some low-lying weather system spanning from the Welsh borders to the east coast, trapped by the jet stream refusing to budge. Sheffield, and South Yorkshire more generally, bore the brunt. Around me even the streets were in spate, impromptu rapids streaming through the piles of plane tree leaves.

I had been given some warning that the water table was rising beyond anything we had experienced a few weeks previously when I accidentally drilled through a ground-floor pipe while attempting to screw up a coat rack. I sprinted down to the cellar to turn off the stopcock on the fountain cascading from above and noticed the lowest point of the cellar where we keep our water tank was flooded by a few centimetres. I presumed it had been from the pipe and when the water had receded by the following morning gave it little more thought.

On 7 November, the sky darkened to a shade depicted by Turner, and the already near-constant rainfall started to increase in intensity. I was writing at home that day, and watched the rain hammer away the last leaves clinging to the hornbeams and field maples on my street and heard the squelching steps of people hurrying home through puddles. By lunchtime it was already nearly dark and, fed up of scrolling through Twitter on storm watch, I felt a sudden need for human interaction. Boots on, hood up, I sloshed down the street to my nearest sandwich shop, which was standing room only with people peering out at the rain through steamed-up windows. The owner told me her cellar had already flooded but she had managed to move the stock out in time. She poured me a cup of tea out of a shrieking tap and I sat huddled up on a bench watching the pavement boil and feeling oddly comforted by the company of strangers.

When I returned home I checked the cellar again and noticed water had seeped back up through the flagstone floor – quite clear and presumably emanating from some saturated underground spring. I checked a few hours later, by which time the River Don had started to breach its banks in the north of Sheffield and the rain was coming down with renewed ferocity. My cellar was now an inch or so under and it would not be long before it reached the electrics in the bottom of the water tank. I rushed around moving things out of the way to other higher parts of the cellar, waiting for Liz to come home with the car so we could drive to our nearest B&Q. We raced past the Christmas decorations and white plastic trees at the entrance and straight out to the woodyard where we grabbed half a dozen hessian sacks and a couple of bags of builder's sand. The man at the checkout wished us good luck. There were others in the city that night who would need it far more than we did.

Back in the cellar we filled the sandbags and arranged them as best we could. Finally I switched off the pump and immersion heater connected to the tank, and we retreated upstairs to sit out our fate. The rain wasn't forecast to cease until the following day. I checked online and saw pictures of streets in the centre of Sheffield under water while families who had gone to the nearby Meadowhall shopping centre – a retail monolith that is nicknamed locally 'Meadowhell' – to watch the Christmas lights switch on had been trapped inside after the Don had breached its banks. Pictures posted online from inside Meadowhall showed groups of delighted teenagers posing in new Primark pyjamas preparing to bed down for the night while some tearful parents posted video messages saying they did not know when they were going to be able to make it home. Another photograph of a riverside pub in Kelham Island showed the River Don lapping the keel of a galleon that had been graffitied high up on the redbrick embankment. That meant the river was back at the height it was during the disastrous 2007 floods which claimed two lives and caused millions of pounds of damage in the centre of Sheffield. As we went to bed, the Environment Agency was beginning to issue red warnings, meaning there was once more a threat to life in our city.

The following morning our domestic barricades had more or less held, although it was clear we were going to need an electrician to repair the damp wiring and to rebuild part of the cellar before the next downpour should strike.

We were among the lucky ones that morning but the anxious night of waiting had for the first time left me experiencing a fraction of the distress that all those flooding victims I had interviewed over the years described to me: the sheer vulnerability of waiting for your home to be invaded, inch by inch, and knowing there is nothing you can do about it.

After breakfast I walked down to the River Sheaf, a few minutes from our house. Viewed from a bridge by a block of new-build flats, the roiling brown river had stained its brick embankments several metres higher than usual but was still within its limits. A couple of others had also gathered on the bridge to watch. Our neighbourhood was safe, we smiled, relieved, not even considering the impact further downstream.

One of the victims in 2007 was a 14-year-old boy who drowned in the Sheaf, not far from where I was standing, after being swept away in a local park where the river runs through. I regularly jog along its banks and have often thought about what occurred that fateful day. Normally the Sheaf barely reaches the knees of the herons that stand fishing in the park, and I often spot grey wagtails dancing over shallow stepping stones. Standing on that bridge, I finally comprehended how the river could so suddenly transform into a roar.

In the city centre stands a memorial to another victim of 2007, a 68-year-old man who drowned after becoming stranded in his car. Since then tens of millions have been spent on defences along the River Don, where the Sheaf, Loxley, Rivelin and Porter Brook all converge. That November when the monthly average of 84mm of rain fell in 36 hours provided the first real test and, largely, these defences held. Pocket parks (where the formerly culverted riverbanks have been opened up into green spaces, allowing the river to tactically flood) and sustainable urban drainage schemes installed along the Don served their purpose. In the lower Don valley the newly constructed concrete walls, barriers and six hundred or so one-way drain valves protected the city's manufacturing heart. Sheffield was spared.

But all that water racing off the moors and through the city's rivers had to end up somewhere. By the cold economics through which flood-defence priorities are calculated, as a main urban conurbation Sheffield was deemed more of a priority than the smaller towns and villages downstream. Ultimately, somebody, somewhere, had decided we were worth more to preserve.

Similar calculations would have been made four centuries ago when, in 1626, King Charles I commissioned the Dutch engineer Cornelius Vermuyden to drain the marshy lowlands of South Yorkshire, North Nottinghamshire and North Lincolnshire known as Hatfield Chase to create new swathes of agricultural land. Within eighteen months Vermuyden had changed the course of the River Don. An estimated 12,400 hectares of fenland was reclaimed during the work and ownership was divided into three parts between Vermuyden, the Crown and the existing tenants who claimed Right of Common over the Chase.

The manor of Fishlake was among the land holdings granted to Vermuyden. Established on what was then the banks of the River Don, the village takes its name from the Old Norse meaning 'fish stream'. Residents had grown wealthy harvesting river fish and eels and establishing ferrying services and a port along the Don. Its fine medieval church, which dates back to the twelfth century, is testament to this former wealth. The church is named after St Cuthbert, he of the aforementioned eider duck and the Lindisfarne hermit who lived in isolation surrounded by water on the Northumbrian isle of Inner Farne. After he died, his body was supposedly carried through Fishlake as it was transported through the country en route to its final resting place in Durham Cathedral.

Situated in low-lying countryside, close to several major rivers and with a predominantly clay soil, the area had always been susceptible to flooding, but land drainage was effectively maintained by a complex system of medieval field dykes still apparent today. According to the Fishlake History Society, there are numerous references in the pre-1627 accounts in the parish bylaw book to the land being 'gripped and groited': cut through with gullys to improve surface drainage. Vermuyden and his fellow Dutch engineers violently disrupted this centuries-old balance. Rerouting the River Don eventually took it south of Fishlake and stripped the floodplains of their ability to store water. In time, new meaning would be attached to the village's historic name.

The Don runs about 30 miles from Sheffield to Fishlake and in November 2019, during the continued rains over the course of the next few days, it was here that the worst of the flooding occurred. I visited on Monday, two days after the Don had burst its banks and engulfed the village. I wanted to talk to residents about their own experiences of an event that, in part, my city had caused.

Fishlake remained all but sealed off to the outside world. To even get close I had to ignore two road-closure signs and cross a bridge with the swollen River Don lapping at its sides. I held my breath as I drove across. On the previous Friday a woman had died in Matlock in Derbyshire after her car became trapped in a ford.

After crossing the river I soon met a police roadblock. Two officers waiting in a car told me there was no way in. I tried another route via a lane called Fishlake Nab but that road soon became impassable. Across the horizon I saw a pair of Environment Agency officials in wellies trudging towards me. They said they had come from Fishlake on foot but it had taken three hours. Looking at my cheap shin-high wellies which I normally only use to muck out the hens at the bottom of my garden, they said I didn't stand a chance.

Back at the roadblock I skulked around to see if any agricultural vehicles might be coming past who would give me a lift and eventually a pick-up pulled over. The driver promised to take me in as far as he could. Nodding at the police officers he told me to put on my seatbelt as we drove into the flood. 'The bastards will pull us over even when the road is shut,' he said.

His name, I discovered as we rattled along, the wake from his tyres coursing into the hedgerows, was Charlie Watson. He was fifty and had lived around Fishlake his entire life. He told me he had never seen anything like this before. His brother had been flooded out and he was heading towards a farm to check on another relative.

Eventually we reached a point so deep there was no chance even his vehicle could have made it any further.

About 20 feet away I could make out the rear wing of a white saloon car that had been entirely submerged. Charlie told me my best chance was to wait here for a passing tractor to give me a lift.

He drove off on a side road towards his relative's house, leaving me standing on the only bit of asphalt poking out of the water. The landscape was as if a world had turned upside down. A great clamour of rooks and crows swirled through the leaden sky surveying the shimmering lake that had displaced the fields of winter stubble. Nearby a kestrel hung with furious patience in the air, scanning this new waterland for some sign of life.

After a quarter of an hour or so, a tractor sluiced towards me, driven by a stern-faced older man. I shouted over the engine noise that I was a journalist attempting to get into Fishlake. There was no way in from here, he told me, better head to the north side of the village where people bringing in supplies had managed to open up a route. I trudged back along the road I had come down, a 30-minute hike in wellies that had already started to leak. This is another rule of writing about floods I neglected to mention: you always get your feet wet.

Following the farmer's directions, I headed back over a canal before eventually coming to a narrow lane that led to Fishlake. The water here was several inches deep, just shallow enough for a car to get through. I drove in past the Fishlake Cricket and Bowling Club, whose lawns now rippled with water, and parked on a bit of higher land. As I walked into the village I noticed the top of a swing ball post poking out of a drowned front garden next to the bobbing heads of several statues.

In a matter of steps my wellies had entirely filled and the freezing water gnawed at my toes. A man in camouflaged fishing waders, carrying a USB charging point in a bum-bag to help people power their phones, wallowed over. He talked me through what had happened. The Environment Agency had only warned people late on Friday night that there was a serious threat to life and they should evacuate immediately. By then, he said, some were simply unable to leave their homes, while others had decided to remain in the village because they were afraid of looters. He told me people had been sleeping in the local pub. The church, meanwhile, had been turned into a rescue shelter.

Our conversation was interrupted by a man called Peter, a neighbourhood watch officer and one of Fishlake's volunteer flood wardens, who was busy coordinating the next day's arrangement of the long-awaited Environment Agency pumps. His beautiful mid-eighteenth-century home now languished under several feet of water. So far he had been refusing interviews with the journalists making their way into the village but was interested to hear I was writing a weather book and invited me into his wife's office in an outhouse where his family had decamped.

They had power to boil a kettle, linked by a giant extension cord to his son's house nearby, and a wood-burning stove. His house could be reached from the study by four metal garden chairs arranged like stepping stones. I left my wellies, and saturated socks, at the door and sat barefoot by the fire while Peter served tea. I noticed a rifle propped up next to the sofabed where he and his wife were sleeping.

He told me he had lived in the village his whole life, bar a five-year period in exile in Hull, and was only aware of two floods on this scale to have affected Fishlake before: in 1923 and 1947. The speed at which the water came into the village this time, though, was unprecedented. He didn't wish to go into too much detail about the extent of damage he personally had suffered: suffice to say furniture, fittings, electrics and the rest. The family's Labrador, meanwhile, had been loaned out to friends on preference to being marooned in the outhouse. A geographer by training, he put the scale of the damage down to three factors: freak weather, global warming, and our destruction of the resilience of the land. The inability of the moors to hold back water to prevent flooding is the same thing that now makes them so susceptible to wildfire.

'I work from peak rainfall to peak run-off,' he told me, reflecting on his time studying as a geographer and the speed at which rainfall could flood down from the moors into the valleys below. 'When I was a younger lad it took twenty-four hours to run off from the moors. Now my suspicion is it is only twelve hours. It's caused by how we manage the uplands above Sheffield but also through the concrete jungle of the city. Then they exacerbate it by not doing anything about it elsewhere. Working down the river, eventually someone is going to cop it.'

I left his house and waded towards the church, my wellies once more instantly filled with the filthy floodwater. A trailer carrying a group of broadcast journalists, some of whom I recognised from being together on former jobs, drove down the main street. The church bells were still ringing out in defiance of the deluge. Inside, pews were stocked with donations of food, blankets, wellies and bleach that had poured in from the surrounding areas. The Hare and Hounds pub, next door, was similarly buzzing with activity. In the back room, camp beds had been laid out for residents unable to return to their homes.

Here I met a man called Andrew Benford, who had been sleeping in the pub since the early hours of Saturday morning. Shortly after midnight, water started pouring into his bungalow, eventually reaching four and a half feet up the walls. He escaped and also managed to rescue his 87-year-old mother from another part of the village. She had since been staying with relatives in Doncaster, but despite receiving no end of offers he was unwilling to leave as his border terrier, Maisie, who was snoozing on a sofa next to him, seemed settled after the trauma of recent days. 'It hasn't really sunk in at all yet,' he told me. 'I just feel numb really.' Friends had been joking with him about the joys of being stuck in a pub but he told me that when he finished his pint there was nowhere for him to go.

★ ★ ★

That November of 2019 turned out to be the wettest ever in Sheffield, according to records that date back to 1882. By 17 November, 441mm of rain had fallen since the start of the month. The previous record was 425mm, set in November 2000. We celebrated my wife's birthday that day with a family meal in our kitchen as the rain pounded against the glass and our young nieces grew increasingly stir-crazy. Cleaning up after everyone had left, we wondered how much more of this weather we could take.

It reminded me of the Pembrokeshire village of Eglwyswrw, near Cardigan, which during the winter of 2016 endured eighty-four days straight of rain, five short of the record of eighty-nine days set in Scotland in 1923. Fortunately our own city's run at that unwanted record came to a long overdue end. The following morning we woke to the first sunshine for what felt like weeks and the psychological impact was immediate. Eating breakfast I watched blue tits scuttling over the feeders before a fat squirrel wedged itself in the middle of one, sending birdseed scattering out across the lawn. The sky was bright blue and the sun picked out familiar shadows, contours and colours that I felt I had last seen a lifetime ago. My water butts gurgled, filled to the brim.

Was all this rain an anomaly or a sign of things to come? Some people argued neither. I wrote a weather column and a dispatch from Fishlake for the newspaper, which quickly got picked up by some climate-change-denying website, whose contributors argued that regardless of the official figures, that sodden November was nothing out of the ordinary. The scientific basis for this seemed to be nothing more than an old newspaper photograph from Fishlake during a previous flood in 1947. I also received a letter of complaint from a disgruntled reader disputing the Met Office figures that there had indeed been record rainfall. Climate change was not the issue, he assured me, solely a slow-moving weather pattern allied to general poor land management.

Certainly that latter point played its part in Fishlake. The destruction of the sphagnum bogs above Sheffield, giant sponges capable of holding back vast amounts of water, and repeated draining, burning and grazing of land over centuries, has created ideal flooding conditions. A few nature reserves dotted along the Don still act as floodplains when the river bursts its banks, but the rest of the land has been left unable to cope.

These localised catastrophes are finally revealing to us the urgent need to repair the wider landscape and lessen the impacts of future floods. I have visited projects in Pickering in North Yorkshire and Stroud in Gloucestershire where so-called natural flood management techniques have reduced the risk to communities downstream. In Pickering I walked with an Environment Agency expert through Cropton Forest, surveying work within the entire river catchment area, which encompasses 70 square miles, to interrupt the downwards flow from the moorlands above. Hundreds of dams had been put in place by simply felling a few of the birch, pine and alder trees that line the river bank while heather bales have also been placed in the river, helping to avert numerous floods. Elsewhere in Britain, beavers are now being introduced to serve a similarly important purpose.

Admirable – and vital – as this work is, it remains on a small scale compared to the reams of scientific evidence of our rapidly changing climate and weather, which even the most ardent deniers are finding increasingly difficult to ignore. The number of so-called 'extremely wet days' has increased by 17 per cent in the past half-century. That means extended periods of extreme winter rainfall are now seven times more likely. One study into the (previously record-breaking) 2000 November floods found climate change increased the risk in England and Wales by at least 20 per cent, and perhaps as much as 90 per cent. Should that trend continue, then whole communities could soon become unviable. Or thinking to the 'For Sale' signs I had seen outside some of the flooded homes in Wainfleet and Fishlake, perhaps they already are?

The long-term impact such disasters had on people stayed in my mind for long after I had left Fishlake and the village had started to rebuild. On my travels to flooded areas around the country people had explained to me the utter helplessness they felt. In *The Great Flood*, Edward Platt described conversations with some victims saying once their homes had been inundated they actually experienced a strange sense of relief, as if the wait for the inevitable to occur is somehow worse than the reality. These are invisible watermarks left in people's minds long after their plasterwork has dried out.

In Calderdale, my home for three years and the district where I first worked as a journalist, people now grapple with this fear every time it rains in earnest. This is a fluvial landscape. Communities live at the bottom of precipitous hills that rise up from the River Calder as it runs from Yorkshire to Lancashire. The first settlers here lived on the tops, but people were drawn down into the valley floor by the cotton and worsted spinning mills of the textile industry that made the area rich. Daniel Defoe visited in 1724 as part of a nationwide tour. On the way into Halifax, the capital of Calderdale, whose magnificent eighteenth-century Piece Hall was the epicentre of its textile trade, he described 'the houses thicker and villages greater'. Its inhabitants, he noted, 'are people full of business, not a beggar nor an idle person to be seen.' That business has long since gone, and many of the grand municipal buildings it helped pay for are suffering signs of neglect. But the water still flows.

On Boxing Day 2015, residents along this 20-mile stretch of river that runs through the valley, including friends of mine, awoke to the wailing of flood sirens. Thousands of homes and businesses were impacted in the towns of Brighouse, Elland, Todmorden, Sowerby Bridge, Hebden Bridge and Mytholmroyd, the birthplace of the poet Ted Hughes. Economists counted the eventual cost of the damage at £170 million but the personal impact ran far deeper. People who had lived in these communities their entire lives were being forced to accept that they no longer felt safe.

That same day in 2015, theatre director Al Dix was returning to his home in the milltown of Saltaire (a few miles from Calderdale) after spending Christmas in Cambridge. He recalls driving up the A1 under a brilliant blue sky and only being aware of the scale of the floods that evening when he attempted to visit a friend in the next valley for dinner and discovered the roads were shut. I've known Al ever since I lived in Calderdale, longer perhaps. The best friend of my own best friend's dad, he is also another fellow Londoner who fell in love with the Yorkshire landscape as a student. They have a word for such intruders in West Yorkshire: 'ofcumden'. Al has lived here for fifty years or so and still can't shake it off. I wanted to talk to him about how people come to terms with flooding because of a commission he was awarded in the aftermath of the Boxing Day disaster: to create a community opera as a way of bringing the valley together in response to what had occurred.

That the Calder Valley is at risk of flooding is, of course, nothing new. The difference in 2015 was the speed and severity. Every person had their own personal weather memory of previous events and invisible tidemarks where the water might typically reach, but never exceed. Christmas 2015 turned all of that perceived understanding of the landscape on its head. The production was paid for out of some of the disaster-relief funding due to the realisation that simply installing new barriers and concrete defences would not be sufficient this time around. The toxicity of uncertainty had seeped into communities along the valley and the people themselves also needed help building resilience.

Over coffee in his warm kitchen on a damp, lustreless day, Al told me that Calderdale Council and the NHS also dispatched specialist psychological teams to work up and down the valley helping people come to terms with the catastrophe and the fear that it would in all likelihood be repeated. Just two weeks before we met he was running a community session in Halifax when people started receiving Environment Agency flood alerts on their phones and immediately sprinted off to their families. In February 2020 I found myself once more in Hebden Bridge, covering its fourth serious flood in eight years. I was told by one business owner who had been washed out again that in the wake of the 2015 floods a friend of his committed suicide.

Al's production was called *Calderland*. It involved a cast of hundreds of residents and was performed in the open air to a thousand-strong audience over three consecutive evenings. Prior to conducting rehearsals Al's team embarked on a story-gathering exercise throughout the valley, interviewing as many people as possible to glean their experience of the floods. The community choir worked these recollections into song: tales of an old people's home being evacuated and a refuge centre stocked with three thousand crumpets and chicken biryanis.

During this process Al discovered something I had noticed myself: the most damaging aspect is not the immediate aftermath but the isolation that follows when people retreat back indoors to rebuild their lives. He too likens it to a bereavement. 'At first everybody rallies around you and are with you for a time but at some point you end up sitting in a room on your own. And the world has moved on. That's what it does.'

What *Calderland* offered is a chance for the communities affected to shape their own response to the tragedy. The same idea is now being taken to areas on the east coast of Britain that are at risk of being wiped out by coastal erosion, to help people come to terms with the idea that their lives as they know them are coming to an end. 'For a short period of time we put community spirit and the wonderful things people did in the limelight and they were proud of what they had achieved,' Al says. 'People were crying in the choir and in the audience as well. I've never made people cry so much in a show.' One night of the performance, typically, the heavens opened. Rain mingled with tears washed down the faces of the audience.

As well as the opera, the project also involved a community arts scheme in each of the affected towns. In Sowerby Bridge they designed a giant goose, 20ft (6m) high and equipped with a hose that could spit water out, inspired by the famous birds that wander the town. It made its public appearance at the Sowerby Bridge rush-bearing festival, an old ecclesiastical tradition where the rushes that once served as insulation on the earthen floors of churches are collected and fresh ones strewn in their place. I had followed that old rush-bearing cart between pubs and churches on numerous occasions in previous years and loved the idea of the giant goose interjecting into proceedings, spitting out the very water that had previously forced so many out of their homes. For as our weather changes we will need to discover new rituals to come to terms with what we are going to face in the modern era. That goose, to me, is a glimpse of a new tradition inspired by very modern weather events, and a symbol of a community refusing to be cowed.

CHAPTER TWELVE

The Vast Machine

The Royal Hallamshire Hospital is the third-highest building in Sheffield. You can see it from the Peak District; you can see it from the hill at the top of our road. And as the date of our appointment there grew nearer I could feel it looming over me wherever in the city I stood.

On a cold, drizzly early February afternoon we walked up towards the imposing 19-storey concrete monolith. I mindlessly scanned the rain-slicked streets to focus on anything except for what was ahead. As we passed a house near to the hospital I noticed the first yellow flowers of a forsythia already in bloom in the front garden, when, typically, in my garden at least, I would not expect it to emerge for another month. The young petals reminded me of the colour of Easter chicks. As we entered the hospital complex, clouds of steam drifted out of the heating vents and dissipated into the milk-white sky.

We waited in the reception room of the gynaecology department, warm and stuffy in our coats though reluctant to take them off. On the muted television screen screwed into the wall were images of yet more winter floods. The covers of the well-thumbed glossy magazines on the tables glistened greasily under the striplights.

A few other young couples like us were dotted among the plastic chairs, all of us avoiding eye contact, staring instead at the hand-sanitiser station, the sexual health posters and whichever member of staff happened to walk by. Perhaps they felt as I did? The sudden realisation that you are not just an observer of a scene, but part of it. That you may still have your coat on but you are no longer walking through.

After a while we were led into our appointment with the hospital consultant. Our respective medical histories were outlined before us and chances of conceiving naturally were weighed up until we were given our own forecast. The consultant, a gruff, weary older man, reaffirmed broadly what we already knew, that whatever 'complications' there were between our bodies were not easily explicable. Throughout the years of trying we had held on to the important truth, so easily forgotten in matters of fertility, that this was nothing more than our bodies not working as we hoped. But as the consultant drilled into us with his brusque questions we both somehow started to feel at fault for whatever glitch it was that had occurred between us.

Interventions loomed. He mentioned a potential hormone-boosting drug. Should that not work, he warned us, our next fertility option was to embark on a course of IVF, which we knew already would be a step too far. Put simply, we felt too raw already. Liz did not wish to subject her body to it, and neither of us our souls. We left with an undetermined date for a follow-up appointment in six to eight weeks' time, a handful of leaflets, and a shared understanding that we were not at the beginning of a process but somehow nearing the end.

In the process of writing this book I have tried to be wary of generalising something which by its nature is intensely personal, but here are some: infertility makes you question your purpose in life. It forces you to blame yourself, interrogating how all the choices you have made up to that point might have led to the ultimate choice of procreation being taken away. It makes other people feel uneasy. It tests the strength of love and friendship as your lives begin to diverge from those around you. At its worst, the term 'trying' makes you feel as if you have somehow failed.

Trying for us, we hoped, would come to mean something different. We were slowly becoming reconciled to the fact that we might not have children together, but knew that would never mean we would stop trying for each other and whatever life we decided to have. Trying might mean spending more time with our nephew or nieces, or perhaps adoption. Ultimately it was a resolution that the pain of the last few years would only serve to strengthen us. That whatever happened to us, we wouldn't stop trying.

We walked out of the hospital taking in the view of the city where we chose to lay down roots, which in a matter of a few miles morphs from the student tower blocks and offices dominating the city centre to the dark moorland borders of the Peak District. Looking out across the rows of terraced houses stretching into the distance, I longed to be back in the small space we had made for ourselves among them. The old rickety house with the damp basement which we were adapting and improving in our own image, and the patch of garden where we monitored all the life that thrived – and died – around us. The place where we watched the vagaries of the seasons as they passed and excitedly informed each other of tadpoles in the pond, birds nesting in the trees and which of our three chickens had been first to lay an egg that morning. The patch of land we already understood better than any other and which served as a continual reminder there is life beyond that which we can ourselves comprehend. Hand in hand, we headed for home.

★ ★ ★

In the Patrick Kavanagh poem 'Innocence', the Irish poet writes of a life on a small hill farm bounded by whitethorn hedges. In what is a celebration of parochialism he describes how an immediate focus on a familiar landscape can unveil universal truths. At the end of the poem he also reflects on the immortality that such an intimate study can bring. He can only truly die, he writes, should he ever stray beyond the whitethorn.

Home, landscape and weather have long intertwined. Writers of our earliest texts often cast the three in opposition to conjure a sense of strife and a deep nostalgia for the invisible thread that binds them together. Home in our oldest literature is typically portrayed as a place where the weather is gentle, free from the worst ravages of the seasons. In the Old English poem 'The Phoenix', the mythical bird symbolising the death and resurrection of Christ lives in a Garden of Eden devoid of seasonal ills where 'neither rain nor snow can do any harm there' and the fields lie 'blessed and perfect'. In *Beowulf*, the adventurer finds himself condemned to fight through a 'slaughter-stained winter', separated from home by a frozen sea. He dreams of the moment when the weather will permit him to return: 'And winter locked up the ice-bound waves till yet another year came in the court, as still it doth, which ever guards the seasons, and the glory-bright weather. Then winter was scattered, and fair was the bosom of the earth.'

In late January 2012 I stepped off the bustle of Piccadilly and walked through the courtyard of the Royal Academy in London to see this centuries-old link between the seasons, place and home projected on a gargantuan scale. I was accompanying my mum to a much anticipated new David Hockney exhibition, which had been many years in the making, reflecting the artist's love of a few quiet miles of the Yorkshire Wolds and his observations of how the landscape changes with the seasons.

Hockney had moved back to his native Yorkshire from Los Angeles in 2003 after finding that he had lost something under the endless blue California skies. This sense of dislocation between culture and the weather is encapsulated by an oft-quoted line from Hockney's mother, Laura, when she first came out to visit him: 'It's wonderful drying weather,' she supposedly said to her son, 'but nobody ever seems to hang their washing out.'

Having returned to Yorkshire, Hockney sketched, painted and filmed the four seasons as they altered a single landscape in Woldgate Woods, close to his Bridlington home. As his focus narrowed he burrowed deeper into the intricacies of the effects the seasons engender upon the landscape. One year he was absent for the hawthorn flowering in May, a spectacle he furiously resolved never to miss again. Describing the 2011 painting *Spring in Woldgate Woods*, Hockney briefly talked about the intensity of the experience: 'Grasses came up, the first campion flowers, buttercups, dandelions. The greens were building up.' In another interview he gave in 2019 ahead of the opening of an exhibition at the Van Gogh Museum in Amsterdam, Hockney mentioned a Matisse quote he admired on painting the sky: 'Two kilos of blue are bluer than one kilo of blue.' When it comes to green, Hockney insisted, 'it must be three kilos.'

I was captivated by the colours of his giant canvases and remember walking out on to the London streets afterwards, awash with winter grey, and feeling refreshed having been hauled through the four seasons. Following my own retreat to Yorkshire a few years later I resolved to take a similar weekly walk through my local woods to see the seasons pass through them. Over the years spent watching the first electric yellow brimstone butterflies on the verges in spring, the fox holes heavy with summer scent, crunching over the autumn beechmast and cutting the boughs of holly bushes bearing red berries to decorate our home at Christmas, I have come to understand the virtue of studying a fixed point to make wider sense of the weather. In doing so I have discovered a simple and eternal pleasure derived from any intimate study of a landscape. It hasn't changed much, Hockney once said of Woldgate Woods, but it changes all the time.

This focus on the local underpinned our earliest forecasts, something that is apparent in the precise dialects once used to describe the weather immediately above our heads and the prognosticators and folklore people developed in an attempt to predict how it might change. Such vocabulary fostered a sense of the weather being something unique and personal, from the sea frets known as 'haars' which drifted across northern Britain from the North Sea to the specific Welsh belief mentioned by the folklorist Marie Trevelyan that oak trees can predict the weather as 'the curling of its leaves foretokened heat'.

Today's sprawling computing systems are made up of these individual observations of the weather, which together combine to make something grand and magnificent. As human society is a concept forged from our own individual preoccupations, so I think of the home weather station affixed to my garage roof and know that while its readings are of interest only to those sharing my immediate postcode, at the same time it is part of something far bigger.

★ ★ ★

Like many Victorian intellectuals, the critic, reformer and writer John Ruskin made his own detailed observations of the weather spanning the decades. Living through an era in which weather diaries had exploded in popularity, Ruskin was among the first to realise the potential of harnessing this disparate network of amateur recorders. Writing an undergraduate essay entitled 'Remarks on the State of Meteorological Science' in 1839, Ruskin imagined building 'perfect systems of methodical and simultaneous observations'. He described his vision as a 'vast machine' where the solitary weather watcher might 'find himself a part of one mighty mind – a ray of light entering into one vast eye'. After all, Ruskin noted, 'the meteorologist is impotent if alone'.

At the time of writing Ruskin was still a student at Oxford University and developing the ideas of art, beauty and social reform to which he would dedicate his work. Improving the lot of the city of Sheffield, a place he once described as 'the dolorous city of the dirty Don', was where he devoted a lot of his energy.

Even decades after writing his essay, Ruskin's claim of being able to forecast the weather remained fanciful to many. In 1854, when the Irish politician John Ball suggested to the House of Commons that one day 'we might know in this metropolis the condition of the weather twenty-four hours beforehand,' he was laughed out of the chamber. As with many innovations in human history, it would take disaster to strike before Ruskin's dream of a network of weather forecasts started to become a reality.

On the night of 24 October 1859, a slow-moving storm depression was noted by observers in the Bay of Biscay near Cape Finisterre. The following evening, when the weather system barrelled over from the Irish Sea, it proved the beginning of the worst storm to batter Britain's west coast during the whole of the nineteenth century. By midnight the storm had reached near hurricane proportions with wind speeds of 100mph recorded. An estimated 133 ships were sunk, with a further ninety badly damaged. In total eight hundred lives were lost at sea that night, twice as many as around the entirety of the British Isles over the course of the previous year.

The largest of the stricken vessels was the steamer *Royal Charter*. The ship was en route to Liverpool from Australia and many of those on board were reputedly prospectors returning with newfound riches discovered from the country's gold fields. Despite worsening conditions that evening, the ship's captain decided to press ahead for Liverpool rather than seek shelter at the nearby port of Holyhead. Off the coast of Anglesey, near Point Lynas, the captain signalled to Liverpool pilot boat number 11 to help bring them in, but the storm had grown so severe it could not reach them. Instead they were forced to drop two anchors at the bow and stern and pray they might hold.

In the early hours of the morning both anchor chains snapped clean off and the *Royal Charter* was driven on to rocks with such force that she broke in two. More than four hundred lives were recorded lost – the highest death toll of any shipwreck on the Welsh coast to this day. There are stories of passengers leaping overboard, their pockets weighed down by Australian gold.

The tragedy focused national attention on the need to develop a proper system of weather forecasting and, in particular, storm warnings. Robert FitzRoy, previously captain of the Royal Navy ship HMS *Beagle*, which had transported Darwin during his famous exploratory voyage to Tierra del Fuego in the 1830s, was at the time of the storm already working to develop a new meteorological office that might serve the nation. FitzRoy had been developing a system that could collect weather observations from around Britain's coastline. In the aftermath of the *Royal Charter* disaster he produced a detailed analysis which, he argued, demonstrated that the path of the storm could have been predicted before disaster struck.

FitzRoy was appointed by the government to establish a new storm-warning service utilising the nascent electric telegraph technology rapidly becoming available. At the same time that he was working on his new weather-warning system, Darwin published *On the Origin of Species*, in which he outlined his evolutionary ideas. FitzRoy, a devout Christian proud of his aristocratic lineage, was fiercely critical of Darwin's claim that he might somehow be the descendant of apes. Once close companions, the pair had drifted apart. As Peter Moore notes in his book *The Weather Experiment*, just as Darwin sought to explain the past, FitzRoy attempted to outline the future.

The first of FitzRoy's weather reports was produced on 3 September 1860, noting pressure, temperature, wind direction, speed and proliferation of cloud. FitzRoy intended for the report to be produced each morning at 9am, six days a week, from weather stations at Greenock, Hull, Yarmouth, London, Dunkirk, Portsmouth, Plymouth, Cherbourg, Le Havre, Jersey and Brest. The first storm warning was issued the following year, on 5 February, using a combination of drums and cones hoisted to warn ships both in harbour and along the coast of an approaching gale.

FitzRoy himself coined the word 'forecast'. 'Prophecies and predictions they are not,' he wrote. 'The term forecast is strictly applicable to such an opinion as is the result of scientific combination and calculation.'

By the summer of 1861, FitzRoy published the first Public Weather Forecast in *The Times* newspaper. His conclusions were limited to 'fine, fair, rainy or stormy' but despite the simplicity of the readings and the fact they often proved wrong, they were wildly popular among Victorian society. FitzRoy became anointed the nation's 'clerk of the weather' by the satirical *Punch* magazine. When the accuracy of his forecasts were tested – as they often were – FitzRoy would adopt the title himself to respond in the letters pages of the newspaper.

In 1863 he published *The Weather Book*, intended as a practical manual of meteorology from his years spent at sea. But gradually FitzRoy became exhausted by his critics, and his work. An old depression which had stalked him earlier in life returned and he left his London home for a period of rest at Norwood, south of the capital. On 30 April 1865, he took his own life at home. The final forecast of his lifetime, writes Peter Moore, was published the previous day. It predicted a storm over London.

Towards the end of the nineteenth century, weather stations were beginning to be installed all over Britain and a network for decoding and forecasting the weather was slowly taking shape. Among the far-flung locations chosen to record these new observations was the summit of Britain's highest mountain, Ben Nevis. In preparation for the opening of the observatory, a meteorologist called Clement Wragge climbed the 4,412ft (1,345m) mountain every day between June and October 1881, taking weather readings while his wife logged comparative observations at sea level at the nearby town of Fort William. Wragge's barometers and thermometers were housed in a stone hut covered with a tarpaulin at the summit. In 1883, following a wave of donations from, among others, Queen Victoria, a proper granite block observatory was built in its place with walls 12ft (3.5m) thick. At the opening ceremony the small crowd that had gathered to mark the occasion shivered in snow two feet deep.

The Ben Nevis observatory stayed open until 1904. The team of weathermen stationed there provided the most comprehensive set of weather data ever amassed on the summit of a British mountain. In summer they played rope quoits on the observatory roof and in winter hurled boulders down the sheer slopes to see if they could do so without catching the sides. Day and night they took hourly readings of rain, wind, fog, hail, ice and whatever else the elements swirled up around them.

Those weathermen witnessed some extraordinary sights from their mountain refuge. In September 1894, a researcher at the station called Charles Thomson Rees Wilson encountered a phenomenon first noticed in the Harz mountains of Germany called a Brocken spectre (or mountain spectre), which is produced when a person stands above the upper surface of a cloud on a slope with the sun behind them, giving the impression of their shadow being illuminated by halos of rainbows. The effect had previously convinced less trained observers that they were being stalked by a mountain Big Foot. Wilson's research at the Ben Nevis station led to the invention of his 'cloud chamber', developed as a way of using water vapour to detect ionising radiation and for which he was awarded the Nobel prize.

Sustaining a weather observatory at such high altitude, carting supplies of tinned food up on horseback and fresh water for when the mountain wells ran dry, proved too ambitious a venture for the Scottish Meteorological Society and the building was closed in 1904 due to a lack of government funding. Over the subsequent century the mountain has done its best to shiver the remnants of the observatory off its back and into the steep granite gullies below, but the bare bones of the dilapidated building remain visible.

I visited a few years back, climbing the mountain in the dead of winter. As we ascended through the low-lying clouds a pair of ravens kronked news of our arrival, their calls reverberating off the granite scree and echoing out among the oak and birch forest in the foothills below. The old Irish word for weather forecaster, *fiachaire*, literally translates to 'raven-watcher'. It is a word imbued with a sense of the dark arts.

Near the summit we spotted snow buntings darting among the lichen-splattered rocks. And then suddenly all life was snuffed out as a heavy mist wreathed its way around us. The navigation cairns that lead walkers away from peril on the final approach also disappeared in the shroud.

Fortunately, I was walking with a guide, David Buckett, who knows this mountain intimately. I stayed close to him as we scrambled up to the summit where through the freezing fog I could just make out the ruins of the observatory. David pointed out to me a precipitous drop known as 'gardyloo gully'. It was, he told me, where the weathermen used to throw their chamber pots away.

It is strange how weather sometimes transports you. Standing on the Scottish mountain with my teeth chattering and crampons on my feet, the ruins of the weather observatory took me back to a time in my early twenties when I was backpacking alone through Africa. I had ended up on a beautiful island off the north coast of Mozambique called Ilha de Moçambique. The island was a peaceful and melancholy place with a dark history of colonialism and slavery. I woke each day with the call to prayer and wandered the warren of sandy streets, looking at the crumbling grand old buildings of the former Portuguese administration.

One morning I decided to join a dhow that was sailing a French family of tourists across to a nearby uninhabited island. On it was a centuries-old lighthouse, presumably abandoned by the Portuguese as Mozambique rose up and gained its independence, and left – like so much else – to slowly crumble. I crunched over broken glass in my flipflops into the room where they had once made the weather readings. It was filled with reams of yellowed paper scattered by the wind. Each page was carefully lined into days and hours and annotated in fountain pen drawn in a beautifully neat hand. I was tempted to take one of the pieces of paper with me but thought better at the final moment and left it on the desk. Removing even a single day of the history of this building felt somehow like a sacrilegious act. The lighthouse, and the lives and work of its former keepers, seemed better left as they were. Just as with Ben Nevis, it served as a reminder, I suppose, that regardless of our attempts to understand and categorise it, the weather will always outlast us.

★ ★ ★

Despite the nineteenth century having witnessed the steady realisation of the 'vast machine' forecasting the weather that he had once envisaged as a hopeful student, John Ruskin had at the same time started to suffer from severe bouts of mental illness. A man who had once written about the weather with such childlike glee, of 'bottling skies' and clouds as 'God's daily handiwork', believed in his later years that he had started to notice an ominous change.

In his student essay Ruskin had written how it was the privilege and duty of the meteorologist 'to measure the power, direction and duration of mysterious and invisible influences, and to assign constant and regular periods to the seed-time and harvest, cold and heat, summer and winter, and day and night, which we know shall not cease, till the universe be no more.' And now, as he watched the weather, he forecast an ill wind blowing towards us.

On a midsummer morning in 1871 in Matlock, Derbyshire, Ruskin sat down to begin a letter in what he called the 'dismallest light that I ever yet wrote by'. The sky he described as 'covered in grey cloud – not rain cloud, but a dry black veil which no ray of sunshine can pierce.' Ruskin had witnessed similarly dank weather all through that spring, in London and in Oxford where he was working at the university as Slade Professor of Fine Art. This was the first hint of the 'plague wind' that came to define his later weather writings. Ruskin asked whether the cloud he saw squatting over him like a toad was derived from the poisonous smoke billowing from the proliferation of mill chimneys situated around Derbyshire, or indeed from 'dead men's souls'.

Ruskin elaborated on his doom-laden theme during a series of lectures in February 1884 entitled 'The Storm Cloud of the 19th Century'. His first public talk, given when he was eighteen, was on the colour and formation of alpine clouds. Now here in his late fifties he lamented how those relatively predictable patterns had been replaced by a hellscape of weather where the seasons could no longer be trusted to do as they once did. Ruskin assured his audience that the eighth decade of the nineteenth century would be recognised in history 'as one of phenomena hitherto unrecorded in the courses of nature, and characterised pre-eminently by the almost ceaseless action of this calamitous wind.'

Ruskin's speech was largely derided. One newspaper review written a few days afterwards posited that rather than the climate changing around us, perhaps it was simply the fact that his advanced age had made him 'more sensitive to disagreeable weather'. Perhaps it had. Ruskin died in 1900 having retreated to spend his final years at his home overlooking Coniston Water in the Lake District. His life ended with the century he came to epitomise.

The notion of changes in the weather having a direct impact on human health remained an unproven yet widely held belief in the late nineteenth century. In 1880 a serious outbreak of diarrhoea in Ruskin's adopted city of Sheffield led to a spate of deaths, including many young children. The Sheffield Corporation decided to erect a weather station to help forecast future epidemics and in 1882 installed the gleaming new instruments in the grounds of the Weston Park museum on a hill overlooking the city centre.

The station was manned by the then curator of the museum, Elijah Howarth, who initially started taking readings four times a day, at 9am, 12 noon, 4pm and 9pm, and recording the findings in leather-bound ledgers which the museum still keeps in its archives. Howarth's enthusiasm for his task led to him being nicknamed 'Elijah the Prophet' and in the 1890s the Met Office adopted Weston Park as one of its official climatological stations. It remains in use today as one of the longest-running weather stations in the country. On 24 July 2019, the fifty-thousandth row of data was entered into the ledgers of Weston Park weather station.

Alistair McClean is the person currently in charge of maintaining Elijah the Prophet's old ledgers. When he first started volunteering at the weather station in 1997 they still logged the readings from the rain gauges, solar and ground and air frost sensors by hand, exactly as they did in the nineteenth century. Now, says Alistair, readings are taken automatically but he still feels the burden of history upon him. 'It is a big privilege but also it is a lot of responsibility to have that longevity of record behind me,' he tells me over the phone one day. 'I just hope I don't cock it up.'

He says the data for the end of the nineteenth century, around the time John Ruskin was warning of the weather changing as we know it, shows that it was indeed a period of extreme cold. The coldest year ever recorded by the station was in 1892 – where an average annual temperature of 7.8°C was logged for the year – while the coldest day ever was on 6 January 1894 when an average of -9.2°C was logged. The following year, in February 1895, the station recorded the lowest ever temperature: -14.6°C. At the time there was a fountain in Weston Park, Alistair says, which would regularly freeze solid. Meanwhile, during the winter of 2019/20, he says, not a single record of settled snow was recorded by the station.

The Weston Park weather station records show that things are getting warmer and with higher spikes of extreme rainfall. According to Alistair, there have been more rainfall events of 50mm upwards in the past forty years than there were in the previous hundred. That November in 2019 when Sheffield flooded broke the monthly record set in 2000 of 425.2mm, with twelve days of the month to spare. The following February was also the wettest on record. As well as 2019 being Sheffield's wettest year on record, in July it also logged its hottest ever day. On July 25 that year the Weston Park weather station registered a temperature of 35.1°C on its 50,0001st day of recording.

Events conspired so Alistair and I could not speak in person. In early 2020 a plague wind blew across these isles and Britain – as with more or less the entire globe – went into lockdown. The Weston Park museum closed its doors and when we were due to meet Alistair himself was forced to self-isolate after a family member suffered mild corona-virus symptoms.

As the country came to a standstill and hospitals moved on to a war footing, we received a letter from the Royal Hallamshire saying our follow-up appointment had been postponed to a later date. Other couples who were further down the road than us were left in a terrible situation as a nationwide order was issued to NHS and private fertility clinics to stop treating patients in the middle of a course of IVF, leaving thousands fearing they might have lost their last hope of having a baby. Elsewhere in the country, cancer patients stopped receiving treatment and all but urgent operations were postponed. Our story was a tiny drop in a far greater swirl of cancelled appointments and treatments across the NHS as hospitals braced themselves for the onslaught of Covid-19 – but it hurt all the same to be facing yet more delays at an age when time was not on our side.

One day during the early weeks of the lockdown I read that inner city parks in parts of the country were also beginning to close to the public, and though Sheffield's remained open I wanted to at least visit the weather station in Weston Park before it might prove too late. On the sort of auspicious spring morning that characterised the onset of the lockdown, I cycled up through deserted streets towards the weather station.

Weston Park sits just behind the Royal Hallamshire and my route followed the same path we had walked to the hospital a month or so earlier. I passed the forsythia which was now starting to wilt after its premature bloom, dropping petals on to the pavement. Patients and staff in face masks walked on the streets outside. Someone had chalked a rainbow on to the road – the symbol chosen in tribute to all those working in the fight against Covid-19.

Alistair had told me that the weather station had moved three times in its existence. The original location was in front of the steps leading up to the museum but in 1920 it was relocated 20 or so metres away. In 1951 it was moved to its present position to make way for a conservatory installed to mark the post-war Festival of Britain.

At the park I propped my bike against the wrought-iron fence that protects the weather station in lieu of the wardens who once kept order here in Elijah the Prophet's day. I could make out its Stevenson screen – a louvre-vented box on stilts designed to house the instruments inside, the rain gauge and temperature sensors on the grass. On the roof of the museum, Alistair had said, was an anemometer measuring wind speed, a pyrometer measuring solar radiation and a device recording the hours of sunshine received in any given day. The museum also still keeps something called a Campbell-Stokes sunshine recorder, which was installed in 1896 and works by focusing the rays of the sun on a small piece of card which is specially treated to produce a burn mark that logs the sun as it rises and sets (though today it isn't used).

I sat on a nearby bench looking out across the park, which was largely empty except for a few families messing about on the lawns. Instead of traffic trundling past there was utter silence. I heard the distant call of a pair of goldfinches gradually growing louder until they came into view and headed off in the direction of the hospital behind me.

I thought of the measuring devices and all the years of weather they had recorded here: of the seasons as they passed through two world wars, the Spanish flu pandemic and now this coronavirus outbreak. I thought of the millions to have lived and died here over that time and how against this sprawling historical backdrop the story of two people hoping to bring another life into this city seemed insignificant. And yet it could not have been more significant for us. At that moment I cried in the dappled shadows where I sat for a life that didn't even exist, except in our hopes.

I'm not sure how long I remained there but after a while I noticed a young boy of about twelve or so had run over to my bike, which I had left unlocked against the weather-station gates. After inspecting it for a while he called out to his mother, concerned it had been abandoned. He hadn't seen me, this invisible figure folded into the shadows, but I quickly gathered myself and shouted over to him not to worry and that the bike was mine. He glanced at me, smiled and scampered off out of sight.

CHAPTER THIRTEEN

Weather Notes

Robin Hood's Bay is a place of rare beauty, as shaped by the weather as the smooth stones that roll in off the North Sea and collect on its shore. Backed by the wild heather expanse of the North York Moors it opens up towards the ocean (in the words of the nineteenth-century master mariner Jacob Storm) 'like a theatre'. I like this idea of the roiling sea as the stage and the bay's inhabitants the audience, braced for whatever wild display is conjured up for them.

My dad's side of the family comes from the North York Moors, and Robin Hood's Bay is a village and coastline I know very well; even the shape itself is traced out in my mind. Northwards beyond the jutting headland known as the Ness is Whitby harbour and beyond that the golden expanse of Sandsend. I have walked the length of these cliffs and motored underneath them by boat watching gannets dive and the waves curl.

Despite visiting Robin Hood's Bay on many occasions in my life, I only first heard about the Storms a couple of years ago. At the time I had a freelance gig writing features for the annual magazine of the Yorkshire tourist board and my commission that year was to interview a brewery owner whose new range of beers was themed on the smuggling history of the bay. It was a ropey-sounding feature but one I was delighted to accept, mainly for a chance to revisit Robin Hood's Bay as a night's free accommodation was included. I had a photographer in tow and the brewer arranged for a friend of his to dress up as an old sea captain of dubious historical provenance, complete with frilly lace sleeves and a three-cornered hat. We trooped down towards the seafront like a disconsolate carnival parade, inspecting the former smuggler hideouts and pretending to ignore the stares of day-trippers around us.

We finished up near the top of the village where the photographer wanted to take a wide photograph of the bay. Waiting while he snapped away, I wandered up to a grassy bank with a wooden bench looking out to sea. On it was an inscription: 'In memory of the Storm families who have lived here since Tudor times.'

I was immediately fascinated by this link between weather and place and curious to know more of the history of the Storms. I asked my guide if there were any family members still living here that he might be able to put me in touch with. Fortunately he was a kind soul and similarly interested in local history, and did not seem to mind my rapid diversion away from questions about his beer.

He said he could point me towards numerous headstones in the village graveyards but as far as he knew the modern-day descendants of the Storms had all moved away. Later he emailed me a history of the family and village written by a man called Alan Storm, which had been submitted for a degree as a doctor of philosophy in 1991 and dedicated to the memory of Raymond Storm, 'master mariner 1892–1971'. I read it entranced: the story of a family and community defined by the precarious existence on the edge of the North Sea and a name shaped by the weather which, as I have seen here with my own eyes, can turn in a matter of seconds into the most ferocious gales.

The Storms, I later discovered, trace their lineage back to 1539 when a survey of Whitby Abbey properties during Henry VIII's Dissolution of the Monasteries named John, Matthew, Peter, William, Robert and Bartholomew Storm as tenants living in Robin Hood's Bay, though the roots of the seafaring surname are thought to lie deeper. Its origins can be gleaned from the *Orkneyinga Saga* written around 1200 which tell the histories of the Viking earls who once ruled the Orkney Isles.

The opening words of the book speak of a king called Fornjótr whose three sons are named Ægir (sea), Logi (flame) and Kári (wind). The descendants of the three sons each bear their own names, channelling the awesome power of the elements. Kári has a son called Frosti (frost), who in turn has a son, Snaer (snow). When he becomes King Snaer, he establishes his own dynasty with three sons whose names translate as snowdrift, snowstorm and fresh powdery snow. Taking their names from the worst of winter was a badge of honour, conveying a sense that they could weather whatever might be thrown at them. Several variations of these weather surnames remain in use today, but it seems the ones that signify the hardiest of conditions have lasted the centuries: Frost, Snow, Flood and Fogg.

The influence of the Vikings on this part of North Yorkshire is apparent throughout the parish of Fylingdales in which Robin Hood's Bay is situated. Normanby and

Whitby are two of many such examples, where the 'by' denotes a farmstead or settlement. Inland from Robin Hood's Bay stands St Stephen's Church, built in 1822 but on a site where a church has been for a thousand years. Generations of Robin Hood's Bay mariners are buried here, close to a water source called Cross Keld, the Old Norse name for a spring.

According to the memoirs of Jacob Storm, a master mariner who lived between 1837 and 1926, their family name was commonly believed to be of Scandinavian origin, due to its strong connection with seafaring. Once while on a visit to southern Jutland in Denmark, Jacob came across a shop with his own name written above the doorway.

I pursued various means of tracking down modern-day Storms but initially to no avail. In the process I discovered tantalising clues of their lives and many tragedies of generations of young Storms. One photograph retrieved from the Hulton Archive, taken in 1895, depicts four male Storms in the sort of knitted jumpers whose distinctive patterns could be used to identify which village drowned sailors hailed from when their bodies washed up on the coast. In the photograph the men are unloading lobster pots from a fishing coble bearing the name *Gratitude*. According to the caption they are Thomas, Reuben, Matthew and Isaac Storm, a farmer who was also coxswain of Robin Hood Bay's lifeboat, accompanied by a Jack Russell called Spy.

I started to obsess a little about the Storms. I mentioned their story, or the bits of it I had been able to piece together, in several *Telegraph* weather columns, hoping a reader might get in touch who had a link to the family, but nobody did. Then one day, vainly scrolling through online search pages, I happened across a local history website compiled in the early days of the internet by a man called Roy Storm.

It contained a rich family history complete with pages of photographs spanning centuries of Storms who went to sea as fishermen or with the Royal and Merchant Navy. I read of a stern-faced young man called Richard Storm who survived two torpedo attacks in the First World War and the Murmansk convoys of the Second World War, and later took part in the Normandy invasion. And I read, too, of others who were lost at sea. In 1942 Captain William Storm went down with his ship, the SS *Widestone*, after it was torpedoed by German U-Boats during the Atlantic convoys about 500 miles south-east of Cape Farewell. In 1886, Captain Andrew Storm was master of the brig *Magnet*, when it went down off the Suffolk coast with the loss of all hands except one. A register of missing Robin Hood's Bay seamen counts forty-nine Storms lost to the waves between 1686 and 1943.

I emailed Roy Storm but instead received a reply from his widow, Anne. Roy had died two years previously at the age of ninety-one, she told me, but added that the following month she was travelling up to Robin Hood's Bay from her Wiltshire home for her annual visit, and would be delighted to meet there and talk me through the history she knew.

Her planned holiday turned out to be the week after Boris Johnson announced the country entering into a state of enforced lockdown. Anne is sixteen years her husband's junior but still fits into the category of people deemed most at risk of catching the coronavirus. Confined to our respective homes we instead arranged a phone call on the morning we were due to meet in the lobby of the Victoria Hotel, perched on a hillside overlooking the village. After chatting over tea we were to have wandered down through the warren of narrow cobbled alleys to the shore. I was hoping to buy us both an ice cream from the van that is always precariously parked on the sand, looking as if a rogue wave might sweep it away. As we chatted through the history of her adopted surname we took an imaginary stroll along the winding streets we both knew so well and breathed deep the tang of salt spray which even the mention of her name evokes in my mind.

Anne told me about meeting Roy in London in the late 1960s. He was an intriguing man, beyond the unusual surname. His father, Raymond, served in the Merchant Navy and his great-grandfather was a sea captain. Roy had read agriculture at Oxford University during the Second World War but in 1945 broke off his studies to join the Royal Marines. He was sent overseas to India on officer training and also served in Egypt where, at the age of nineteen, he found himself in charge of a camp of German prisoners of war. After returning to England to finish his degree, Roy travelled overseas again in the early 1950s to work as an inspector for agriculture to the Sudanese government. He came home in 1955 and started work as a management consultant. By the time Anne met him he was employed by the frozen food manufacturer and purveyor of fish fingers, Birds Eye Limited. After the years of travelling a nod, perhaps, to his maritime roots.

Anne had grown up in Australia, and still speaks with a trace of an accent. She had come over to England in 1967 in her early twenties with vague plans of seeing the world. After travelling through Europe with friends she worked as a teacher in London. She recalls living in a bedsit in Baron's Court, teaching in the East End, and marvelling at the sheer dreariness of winter in the capital compared to her native Sydney.

In search of illumination she joined a music club based at Holborn Library and it was here she met Roy. Later they happened to both join a club trip to the champagne town of Épernay. They exchanged numbers, though initially Roy preferred to write letters. He first invited her to accompany him to go and see a film about Mozart in the cinema. They married in South Kensington Registry Office in 1971 and Anne Millar became a Storm. 'I didn't think a great deal about the surname at the time,' Anne tells me. 'Apart from the fact it was related to the weather.'

There is a tragic parallel between the name she took in married life and her own childhood. During the Second World War a storm had claimed the life of Anne's father, Flying Officer Thomas Roberts Millar. Anne was just three weeks old when her father, known as Bob, left to undergo flight training. He sailed to America, Canada and later across the Atlantic to Britain. In January 1944, Millar was transferred to Italy as a bomb aimer with 104 Squadron RAF as part of a Wellington Bomber crew taking part in missions across Europe. Later he was seconded to 31 Squadron of the South African Air Force and stationed at Celone air base, Foggia, where he was promoted to leading bomb aimer. Here he ran a book club with a fellow Australian airman and dropped supplies to partisans behind enemy lines.

His final mission was on 12 October 1944, as part of a squadron of twenty aircraft tasked with dropping supplies to Italian partisans in the high passes and isolated villages of the Ligurian mountains and northern Alps. There were four chosen drop zones that evening with five aircraft designated to arrive at each. The chosen drop zone for Millar's B-24 Liberator Bomber was in the Ligurian mountains north of Genoa.

They took off in the late afternoon and the weather quickly turned, a heavy fog thickened and lightning streaked the darkening sky. A slow front was moving towards them across the western Mediterranean, lying from the Balearics to Nice and north-east across western Switzerland, while a depression started to form south of Nice. A veil of cloud drew over the mountain tops at 10,000ft (3,000m), with banks of cumulonimbus and thunderstorms to the west. The mission was to end in disaster.

Only three of the crews managed to achieve their drops; the rest were abandoned. Of the twenty aircraft involved in the mission, six failed to return. The following morning back at the air base there were forty-eight empty places at the breakfast table. The wreckages of five of the aircraft were later retrieved from where they had crashed in the mountains but the sixth, Liberator KH158, on which Bob was flying as part of an eight-man crew, had vanished without trace.

Over the years Anne has attempted to establish what happened to her father that night and where his aircraft came down. She has traced relatives of the other crew members and spoken with numerous specialist researchers. With little else to go on, she believes the Ligurian sea to the immediate south is the most likely location where her father's aircraft ended up. By crashing into the ocean he would have suffered a similar fate to that which has befallen so many men of the Storm family. Anne still has a letter from her father which he wrote to her on her first birthday in February 1944, the year of his death. She has given me permission to reprint it in full here:

My Dear Daughter,

This is the first time I have written to you and although you are as yet too young to read it perhaps mother will save it up until the time comes when you can read it yourself. In 2 days time it will be your first birthday anniversary – a great event for your parents. My regret is that I cannot personally be there to help you blow out your single candle but believe me lassie I will be there in spirit.

I am writing this from a place called Italy which is far away from our fair land – a place where I would not be by choice so far away separated from a wife & daughter so dear to me. But I am here, precious one, because there is a war on caused by certain people who wished to rule the world harshly & despotically, imperilling an intangible thing called democracy which your mother & I thought all decent people should fight for. You will understand as you grow up what democracy means for us & how it is an ideal way of life which we aspire to put into practice.

All I ask of you, Anne dear is that you stay as sweet as your mother & cling tight to the subtle thing we call Christianity, which has been the core of her way of life & her mother's & mine. I hope that you will love & respect me as I love & respect my father.

That's all young lady. Have a happy birthday – may they all be happy birthdays. I hope to be home again one fine day. In the meantime lots of love to you & to mother

From Dad
Bob Millar

Anne and Roy Storm had four children together. She says it took until the eldest two were at university before she made her first visit to Robin Hood's Bay. They stayed at a hotel in Ravenscar overlooking the bay where the Storms had gathered for a family reunion. She has met numerous other Storms over the years: Roy's older brother, Alan, who I discover is the author of the family history emailed to me by the Robin Hood's Bay brewer. Anne tells me of a cousin, Arnold, who died several decades ago and once wrote a book called *In A Teacup*, and his brother, Jacob Francis Storm, known as Fran, who she describes as a 'font of information' but who also died some years back. Another cousin ran the Victoria Hotel, although it has since moved into new ownership.

The demands of modern life have scattered the Storm dynasty across the country and Anne is not sure if a single family member remains in the village. Still it is a place where she always feels at home and, since his death, close to Roy. Their sons have the surname and her daughter has kept it too, passing it on to her own daughters. It is a place which inspires a feeling she struggles to put into words, but one I recognise from the same places I am bound to by family ties. The weightless pull of history which roots us to a community and landscape; an anchor resisting the groundswell of time.

* * *

The story of the Storms matters to me because it represents something that has been lost. Over the centuries a rich and distinct vocabulary has evolved across the British Isles linking people, place and weather. These weather words possess a delicacy and intensity of focus that pinpoints the experiences of communities and generations that have watched the seasons come and go. They denote the power of the weather to shape us and provide a sense of local identity. And as the seasons themselves grow homogenised as our climate changes, so the very words we once used precisely to describe the weather and its effects upon us are drifting silently into oblivion.

Our national weather forecasts, for example, insist on only ever using three descriptions of rain: 'light, moderate and severe'. And this is despite a rich abundance of regional dialects that describe the exact sort of rain that falls on any given part of the British Isles, be it fluddering, bucketing, lashing or – in Celtic parlance – fiss, meaning a clingy drizzle. The author Melissa Harrison has written beautifully of this country's grudging love of rain and has helped excavate more descriptive words. She cites 'plothering' (used in the Midlands and north-east to describe heavy downpours), 'cow quaker' (a sudden and dramatic rainstorm in May) and 'smirr' (fine, misty rain). While the Inuit, so the apocryphal saying goes, have more than fifty words to describe snow, Harrison writes, the English language far surpasses that in assessing exactly what sort of rain it is descending upon our sodden isle.

As for snow, a project launched by researchers at the University of Glasgow a few years ago to compile a thesaurus of Scots dialect recorded 421 different terms. Among them were 'sneesl' (to begin to rain or snow) and 'skelf' (a large snowflake). 'Spitters' are small wind-driven snowflakes, and 'unbrak' the beginning of a thaw.

In his book, *Landmarks*, Robert Macfarlane has gathered what he describes as a 'word-hoard' of our lost vocabulary, many of which relate to the weather. Within the nine glossaries of his word hoard, an entire section is reserved for mists and fogs from 'roke', an East Anglian term for evening fog rising off the marshes and water meadows, to 'daal'mist' which he describes as a Shetland word for a mist that gathers in valleys overnight and is exhaled when the sun rises. In Dorset, icicles are 'clinkerbells', in Cumbria they are 'shuckles' and in Yorkshire they are 'ickles'. In Sheffield, when it snows, it 'snitters'.

This rich lexicon is one derived from mariners, fishermen, farmers, miners, communities in the furthest corners of the British Isles, whose lives were dependent on the seasons. The modern era has seen a hollowing out of many of those communities and the work that defined them dwindle. As our society has urbanised, the young have been drawn into towns and cities and not returned. With the passing wave of each generation, the language – of the countryside, at least, for in the cities urban slang still evolves with an ever-renewing effervescence – has blunted. We are losing our weather words because simply, sadly, we no longer have need for them.

A couple of years ago, the Met Office launched a call-out for such weather words as part of plans to incorporate regional dialect into its local forecasts. There was a flurry ('blirt' in the old Scots dialect) of publicity but I didn't notice any of the words ever making it into an actual weather forecast.

Instead, in 2014 the Met Office started to attribute human names to storms ahead of each winter. This in itself represented to me an exercise in myopia and a loss of linguistic dexterity. An acceptance somehow that we are no longer the audience looking out to the weather, as Jacob Storm once described in Robin Hood's Bay in the nineteenth century, but rather our gaze has turned inwards, upon ourselves.

It strikes me that as we have lost this subtlety of language to make sense of the weather, our relationship with the seasons has become one of attrition. We hear more these days of phenomena such as 'weather bombs' and 'thunder snow', following in the US tradition of describing extreme weather events in the sort of martial language usually preserved for the defence of the realm.

I remember a few years ago lying awake in the early hours of the morning in a hotel room in Seattle where I had been sent to do an interview, watching the rolling news channels to detract from my West Coast jet lag. A storm was forecast to arrive that day, not the sort of Pacific hurricane that can do so much damage in that part of the world, but a more localised run-of-the-mill weather event that had the potential to bring down power lines and cut electricity to a few thousand homes for a couple of hours.

The forecasters were breathlessly describing it as a 'monster storm' with graphics of the isobars as they coiled and swirled played endlessly on repeat. I wondered then what this crude language said about our changing relationship with the weather. That year record wildfires had raged around Seattle, turning the air toxic, the amount of ash and smoke present equivalent to smoking seven cigarettes a day. Perhaps as our weather spirals beyond our comprehension, treating it as an enemy to be feared rather than something to be revered is the natural outcome?

★ ★ ★

One August morning in 2019, I left my parents' home in north London where I was staying and walked to the tube station to head south across the river. It was rush hour. I squeezed in among the clattering carriages and juddering bodies and emerged thirty minutes later at Bermondsey station – five miles or so underground during which I was told countless times by the recorded announcer at the stations I walked through to 'take care' in the hot weather.

A short walk from the station down a side road into the former leather district of Bermondsey, I came to a brick-built warehouse. I knocked on a heavy wooden door and stood waiting in the street. Four storeys up a shuttered window surrounded by bird feeders swung open and a ruffled head poked out. 'You're here,' shouted Norman Ackroyd in a broad Leeds accent and a tone I couldn't quite unpick as either pleased or irritated. A few minutes later the main door opened with a satisfying clank of bolts and he beckoned me into his studio.

I had come that day to visit a man I, like many others, have long regarded as one of our greatest living weather artists. Norman's abstract paintings and etchings capture an elemental essence of the weather from which modernity has largely cut us adrift. Norman has lived here for four decades, initially moving in with his children aged three and one, at a time when it was far from the gentrified sweep of London it is today. He lives on the top two floors of the old leather warehouse, which is where he paints his watercolours. The bottom two floors are reserved for his oil painting and etching.

When I walk in, the ground-floor studio opens up like Merlin's cave. Norman disappears to find coffee and leaves me to take in my surroundings. There is a mammoth 1900-built printing press engineered in London's East End by Hughes and Kimber. Norman took possession of the machine in 1975 and describes it as probably the best etching press in Europe. Wooden shelves cover the walls cluttered full of jars containing pigments and oils which Norman combines to make the etching inks, grinding them down together until they meddle. A not unpleasant chemical smell hangs in the air and I notice a cardboard box under a sink labelled *CORROSIVE. OXIDIZING AGENT*. Next to it is a fire extinguisher. The tools of his craft, the inner workings of the print man's studio, are what grab my attention at first, but tacked to the walls and hanging up all around us are Norman's famous weatherscapes.

The coffee retrieved, and brewed, we head upstairs and sit on two battered leather chairs. Norman lights a cigarette and blows a delicate stratus of tobacco smoke above our heads and we immediately fall into talking about the weather. I am mindful of the famous Samuel Johnson quote: 'It is commonly observed, that when two Englishmen meet, their first talk is of the weather; they are in haste to tell each other, what each must already know, that it is hot or cold, bright or cloudy, windy or calm.' Listening to Norman's weather memories keeps me enthralled for nearly two hours.

Norman Ackroyd is a contemporary of David Hockney. Both young Yorkshiremen studied together at the Royal College of Art and have remained close ever since. Norman has accompanied Hockney as he has mapped the seasons in Woldgate Woods and describes his friend as 'a wonderful draughtsman'. While Hockney's interest, at least in later life, has focused on mapping the seasons, Norman focuses instead on the weather itself. He is drawn, he tells me gesturing to a map on the wall, to the edges: the coastline around Britain and Ireland where, like Turner lashed to his mast, he travels in pursuit of the weather, and draws as it descends around him. He points to an etching of twilight hanging behind our heads which he drew looking towards the Gower Peninsula at 4.50am. Norman describes what he intended to capture as 'a strange quiet before dawn'. It's not the view he sought to evoke, he told me, 'it's the light and the noise and the smell'. The way the weather inflames our senses.

Norman grew up insulated from the weather by heavy clouds of steam and coal dust in the south Leeds inner-city suburb of Hunslet. He describes it as acres of poorly built and overcrowded back-to-back houses pockmarked by craters left over from the Luftwaffe bombs, sprouting rosebay willowherb. All around them were the looming spires of industry: the Hunslet Engine Company, railway yards and coalfields. 'Everything was powered by steam and coal,' he recalls. 'The atmosphere was disgusting. There were days when you wouldn't see the weather through the smog and chimneys pouring out smoke.'

His father was a butcher and his brother followed him into the family trade, helping out at the city abattoir around the back of Kirkgate Market. Norman remembers him returning from the slaughterhouse one day and telling him that the pigs reared in South Leeds had lungs that were entirely black compared to the animals reared in the countryside.'He said to me:"We've got to get out,"' Norman recalls.

The nearby Yorkshire Dales provided their escape. From a young age Norman and his brother would get up early on Saturdays and run to the station to catch the 4.30am milk train out of Leeds. He would watch his brother fishing for grayling and brown trout on the River Wharfe while he sketched the surface of the water and trees around them. For lunch they would pick mushrooms in the fields and cook them with streaky bacon on a camping stove. That is when he started to become fascinated by the noise in a landscape – or the lack of it – and the way the weather can in an instant change everything.

I remember having a similar revelatory experience myself while a student in Leeds and out cycling in the Dales one day. I was descending from Kidstones Pass towards the village of Kettlewell in weather so foul that my brakes started to slip in the heavy rain. I pulled over and stood in what shelter I could find and watched the rain hammering down on the hills. The Dales is famous for its hidden network of underground channels, mine workings and caves that snake through the valleys and I noticed the rain start to pour out of previously unseen crevices. I realised something then that stayed with me ever since: how the weather can animate a landscape. If that is so, I wondered, what must it do to us?

Later, when at grammar school, Norman Ackroyd would also cycle up to the Dales on weekends with friends – the furthest they ever made it to was the limestone crags of Malham. His favourite time to walk in the hills was during, or immediately after, heavy snowfall. 'That is real weather,' he says. 'The stillness of it all. The challenge is to get the sound and smell of a landscape when it is purely visual.'

Norman started his lifetime journey around the edges of Britain on Yorkshire's east coast. We talk in depth about the coastline we both know and love. The previous week, I tell him, I had been whale-watching on a boat sailing from the fishing village of Staithes and we recorded more than twenty minke whales breaching so close to the boat we could smell the cabbage on their breath that led to them being nicknamed 'stinky minkes' by sailors in the past. Norman describes to me a journey he once took from Saltburn-by-the-Sea southwards right along the Yorkshire coast to the towering chalk cliffs of Flamborough Head, sketching the old spoils heaps and alum and potash mines that have shaped this coast. I show him a photograph of a rainbow over the North Sea horizon I captured on my phone during the Staithes whale trip and he fetches a recent etching of a rainbow he has done, marvelling in the 'perfect geometry' of nature.

Watching the weather at such close proximity over so many years he has, of course, witnessed that geometry become warped by climate change. But Norman insists that despite all the alterations of the weather and the seasons he has seen, he has little time for nostalgia or remembering how things were. 'I don't look back,' he says. 'Really, I don't. I'm interested in what things are like now.'

He explains the purpose behind everything he creates is to produce a 'note' that conveys the exact sensation of whatever weather it was. Norman describes the process of etching as a musical one, with the ink he plays with as varied as the keys of a piano. Writers, he reminds me, grinning, are restricted to just twenty-six letters.

Reflecting long after we have spoken on what this means, I imagine it like a tuning fork rapped off a hard surface and left humming in the air. In that one single note are the shrieking sea birds and the crashing waves and the boiling sea being whipped up by a storm, Norman lying on his stomach sketching a clifftop gannetry while the rain reverberates off his back, or the deafening silence of snowfall.

We leave his studio together and walk down Bermondsey High Street to the gallery where a new private exhibition of his works is being put on display. Here I leave Norman chatting to the curator and head back towards the tube station, past the rushing office workers and the coffee shops and the gleaming spires of London Bridge obscuring the blue sky.

I find myself inspired by Norman's refusal to look back and to focus instead on capturing the immediacy of his experience. How the immersion in his work and art serves as a reminder that even though we are losing – and have lost – so much of the former intimacy of our relationship with the weather, there are links that can never be broken.

I think of my own predilection for nostalgia and how looking back to the past, and in recent years constantly looking ahead trying to create new life, has become a burden on the present. For above all what Norman captures is his own ever-renewing sense of wonder in the weather: the sheer childlike joy he felt in the Yorkshire Dales all those decades ago. He refuses to allow himself to be weighed down by that memory, but instead seeks ways of consistently reinventing it.

Doing so, I realise, is truly living by the seasons. And even as our weather distorts to the point where it threatens rather than sustains new life, the notes that Norman captures will still resonate. The softness of dawn light, the smell after rainfall, the awe-inspiring terror of a storm barrelling towards us. We may lose the words for it, we may lose the very structure of the seasons we introduced to make sense of the weather, but to our very last days on this earth we will never lose that note.

CHAPTER FOURTEEN

Solstice

The midsummer sun appeared at 4.38am, rising as a solid egg yolk slides over a plate. We walked up with it, following a medieval packhorse trail known as the Long Causeway leading out of Sheffield and into the Peak District. Gradually the colours seeped into the surrounding clouds, staining the edges. The violet dawn conjured a strange impression of twilight – a trick of the turning tide when the North Pole tilts at 23.5 degrees to the sun, marking the longest day of the Northern Hemisphere year.

Ahead of us was a local landmark known as Stanage Pole, which stands at what is believed to be an ancient boundary line in the gritstone boulders between the Anglo-Saxon kingdoms of Mercia and Northumbria (and nowadays between Derbyshire and Yorkshire). Visitors have been carving their initials into the rocks at the base of the wooden pole for centuries. The earliest I have ever found date back to the 1600s. That morning two women and a young boy had gathered there to spread out a picnic blanket and watch the sunrise.

Liz and I said, 'Good morning' – noting as we passed out of earshot how that well-worn phrase takes on greater meaning at this moment in the year – and walked on across the silvery cotton grass where curlews and a solitary, sepia kestrel drifted over the bog.

At the edge of the escarpment, a morning mist galloped over the rocks and down into the Hope Valley below. We sat on a flattened boulder, listening to the owls in the woods beneath us hooting across the county borders, and unpacked cold veggie sausage sandwiches wrapped in tinfoil and a thermos of tea. Dotted along the vast ridge of Stanage Edge and into the horizon we saw others gathered in similar acts of midsummer reverence.

During my meeting with Norman Ackroyd at his studio, he had reminded me of Dylan Thomas's 'Poem In October'. Thomas wrote of an early morning walk taken on his thirtieth birthday in an effort to somehow reclaim the seasons of his youth. The weather 'turns' continually in his poem as he reflects on his memories of each season, giving the impression of time slipping through his fingers. Liz and I spoke as we munched our sandwiches of the summers we had spent together and the nights where we had never slept. We called it 'silly o'clock' when we were students, the moments when we briefly cheated time.

The summer after university, not long after Liz and I had first got together, I cycled across northern Europe with two friends. We crossed Holland, Germany and Denmark, heading for a music festival called Roskilde, 20 or so miles west of Copenhagen. The day we were due to arrive happened to coincide with the summer solstice, though we found ourselves still in northern Germany, 100 or so miles away from our destination.

To get there in time we decided to cycle through the night, sleeping only briefly on a grimy and deserted passenger ferry which transported us across the Fehmarn Belt – the short Baltic Sea strait that separates Germany and Denmark. The voyage only took around 45 minutes so we persuaded the ticket officer to allow us to make it three times in order to snatch some sleep on the plastic benches that were screwed into the deck

Eventually he lost patience and when our ferry docked in Denmark for the third time at about 4am we were woken up and ushered off. We started cycling up the only road we could find – a dual carriageway – our rear lights flashing red in the half light. About 45 minutes later the ferry returned, this time bearing the first freight haulage lorries of the morning, which soon roared past, horns beeping furiously. The draught of the passing lorries made us wobble perilously on our pannier-laden bikes as we headed northwards.

We managed to turn off the highway into a village where a bakery had switched on its ovens, wafting a delicious smell of cinnamon and fresh bread across the deserted street. We lay on the grass of a nearby churchyard waiting for it to open, a cuckoo trilling incessantly through the misty morning.

I remember little else about that day, aside from arriving late and wheeling our bikes through the festival site to meet up with friends who had already pitched camp. And I remember vowing with my cycling companions that we would do this every year. Not *this*, necessarily, but always stay up through the summer solstice. We never did again.

That was fifteen years ago, long enough to make me appreciate that time does not flow smoothly. It spits and gurgles and trickles and roars, dashing plans like stricken ships on to the rocks. The earth may continue to orbit the sun and tilt upon its axis each year marking the summer and winter solstice, but experience has tempered the neatness of any narrative arc.

I have written this book – just as Liz and I have lived much of the past few years – with a sense of an ending in sight. We have hoped, at times yearned, for our final chapter to be with a baby in our arms. But it has proved not to be. As I bring this story of the seasons to a close, the 'complications' which resist our efforts continue to do so. We wait, time races, and the world keeps turning.

I also had another ending in mind. To mark the 2020 summer solstice, the first of this decade, I intended to relive the promises of my youth and cycle to Stonehenge to watch the sunrise. I had planned a route to the West Country that would follow, in part, that taken by Edward Thomas as he cycled from London out to the Quantock Hills in the course of writing *In Pursuit of Spring*. This was to be my own pursuit of summer, shared with a crowd of pagan worshippers, drunk teenagers, bemused tourists and newspaper photographers.

After seeing Stonehenge closed during the early days of lockdown I suspected this would prove difficult and, sure enough, come 21 June the solstice, like all else that year, was cancelled. Instead a 30-minute livestream was broadcast on English Heritage Facebook's page of the sun rising over the stone columns. This was not, I'm fairly certain, what my 21-year-old self had in mind about staying up to celebrate midsummer.

I suppose what I hoped to capture standing among the crowd of thirty thousand or so who gather at Stonehenge each year was midsummer as it is observed in the modern day, to understand how those rituals might connect some of the histories of the seasons I had uncovered – how weather, and our response to it, bridges time.

I had a similar experience watching the total solar eclipse in the Faroe Islands in March 2015. As the moon travelled slowly across the sun, the final pools of light disappeared over the Atlantic Ocean and the Faroe Islands dissolved into that brief but immersive period known as 'totality'. The mewing oystercatchers suddenly fell silent and harbour lights twinkled. I looked not out to sea but at the strangers around me, their faces illuminated in perfect rapture by the flashes of their camera phones.

Standing amid the throng of international eclipse chasers, I thought of the Norse warriors who once occupied the Faroe Islands and an old Viking story somebody had told me a few days previously: that the apocalypse would come about from two mythical wolves chasing and eating the moon and sun at the same time. I find such stories often mean little out of context. Watching the world suddenly plunged into darkness, though, I felt an altogether closer connection.

While Stonehenge might, symbolically at least, have been a good place to close, it would also not have been strictly true to this book. For my journey in pursuit of the seasons has taught me more about the intimate connections we develop with them, and how the weather, a collective experience, resonates on a deeply personal level. Had I travelled west, I would have found myself marking the summer solstice alone in a crowd on the other side of the country when the one person I really wished to spend it with would have been asleep in our bed hundreds of miles away back home. And so that midsummer's morning I stayed put and, having persuaded Liz to join me, we set our alarms for 3.45am.

★ ★ ★

Reading about the history of midsummer, I gathered that a walk was in fact a far more historically accurate way of honouring its arrival than gathering at a stone circle. We may nowadays associate the summer solstice with druids and pagan worship but that is dismissed by folklorists as very much a modern interpretation.

The Stonehenge midsummer gathering is based on an assumption for which the primary blame rests upon the seventeenth-century antiquarian John Aubrey. After surveying the site on Salisbury Plain, Aubrey concluded it pre-dated the Vikings, Normans or Romans and therefore must have been the work of druids (the only prehistoric priests mentioned in any classical text). This hypothesis dominated all future interpretations of the site.

We know the stone circle served some ceremonial purpose. Recent excavations have revealed the cremated remains of men, women and children who were deposited at Stonehenge between 3000 and 2500 BC. But the origins and use of the stones remain a mystery, and there is still to this day little proof of an ancient link with the summer solstice.

According to the folklorist Steve Roud, a more exact history of midsummer only reaches as far as the early medieval period, when it was a festival that occurred between St John's Eve on 23 June and St Peter's Day on 29 June. Homes and churches were decorated with birch boughs and wildflowers including St John's wort (which, according to midsummer folklore, if picked by a maiden on midsummer's eve and found fresh the following day indicated she would soon get married). Bonfires were lit, partly for drinking and feasting in the streets, and partly in the belief the smoke helped purify the air – a practice later outlawed by the church for its association with something obliquely termed 'superstitions'.

Another feature of midsummer celebrations were processions which Roud describes as 'spectacular torch-lit affairs'. The Salisbury Museum contains a 12ft (3.5m) figure known as the 'Salisbury Giant' first mentioned in the records of the city's Tailors' Guild in 1570, which was intended to 'sett goinge for the accustomed pageant of Mydsomer feaste'. The giant was accompanied by a wooden hobby horse named HobNob, which cleared the way at the front of the procession.

Even if the two of us walking across the moors on the western edge of Sheffield was a 'procession' in the loosest possible understanding of the word, I still felt an accordance with the midsummer of the medieval calendar. What becomes clear to me reading such accounts is the lack of prescribed celebration about how to mark the solstice, which for many of us persists into the modern day. What spans the centuries, rather, is an ancient human need to rail against the darkness and somehow mark the turning of the year. In order to do that we create our own multitude of rituals.

We saw it heading up to the tops of the moor that midsummer dawn. At viewpoints leading out of Sheffield, people had gathered with cameras on tripods to capture the rising sun. We parked our car at the bottom of Redmires Reservoir and noticed someone had pitched a tent opposite a conifer plantation. At dusk here in early summer, nightjars chur from the forest boundaries. In previous years Liz and I have stood listening to their ethereal calls, which led to old superstitions that they were vampiric birds that fed on the milk of goats by night.

We wondered what other sights we might encounter on our dawn walk. The convergence of the mortal and magical worlds Shakespeare described in *A Midsummer Night's Dream* was in fact based on May Day Eve (in European folklore a time when the boundaries between the two worlds become blurred), though a sense of the supernatural still hangs over the summer solstice. For some it marks a time when other forces come to the fore.

A decade or so ago in the run-up to the summer solstice, there were at least twelve attacks on horses in fields along the border between Derbyshire and South Yorkshire. One horse had eight litres of blood drained from its stomach, others had their tails removed and their manes plaited in intricate patterns. Stones arranged in the shape of diabolic pentagrams were found in nearby fields. In the 1990s a stone circle in the Peak District was also vandalised by a group who had reportedly gathered to worship the devil.

From Stanage we walked across to Burbage Edge where stonechats danced over the bracken tops, their calls reminding me of the clicking shutter of a camera lens. By that stage the bright blue morning had started to cloud over and shadows chased us across the earth. Mindful of the midsummer tradition of gathering birch sticks, I climbed off the path to explore the base of an old silver birch whose calloused trunk bore the scars of many a moorland winter. Near the tree I took one mis-step on the heather and momentarily disappeared into a hole which swallowed me up to my chest. I wondered what the raven who had been carefully watching our progress along the deserted path made of my lumbering movements. As for Liz, I could see her laughing as soon as I hauled myself back out.

Suddenly a cuckoo came into earshot, its song echoing down the valley. Hearing it instantly reminded me of the Danish bird we had listened to while waiting for the bakery to open on that midsummer's morning all those years ago. I thought of what I had learnt in Thetford Forest about how rapidly cuckoos were disappearing from these shores and all that would be lost with them. How a single sound can resonate through the seasons and imbue a landscape with memory. Dylan Thomas wrote of spring and summer 'blooming in the tall tales', well-worn stories of the season which serve to define them as distinct periods in our minds. These are rituals of nature that have developed with the weather, and which in many instances now cannot adapt in time to the rapid changes taking place in the modern age.

As we walked over the moors, looping back round into Sheffield, we heard skylarks bursting into life all around us. The birds, which like the cuckoo are another species in precipitous decline, seemed as vital companions to us then as when they were described in such joyful abundance by Edward Thomas during his pursuit of spring a century ago. Thomas wrote of our capacity for seeing the past in 'rose-colour'. Perhaps one day, he questioned, even his present might be coloured by nostalgia. Listening to the unbowed songs of these diminished birds, I wondered if the same might eventually be said of now?

And then on the final stretch of our walk, came an unexpected encounter. We were tramping through a bog at the time, having veered off from the path in a vain attempt to find the remains of a prehistoric stone circle on the moors where supposedly local pagan worshippers still gather at midsummer. The ground was heavy-going, soaking our boots and draining the last reserves from our already tired legs.

Coming over the crest of a hill we suddenly came face to face with a solitary stag standing just a few metres or so away. We were close enough to meet his unblinking gaze and see the muscles tighten in his haunches as he decided whether to stand his ground. Slowly, and still facing the stag, we retreated, circling away from him. As we did the rest of the herd came into view: another younger and smaller male and three deer – all staring intently at these invaders on the moorland. As we walked away the shapes of the deer gradually blended into the heather and tussocks of grass. We were left with a new midsummer memory, a reminder on the longest day of the thrilling immediacy of the present.

We had no idea when we decided to start trying for a child of where it would take us. The self-help pregnancy guides talk of a 'journey', though that already grating word seems especially so for us, because infertility obscures the final destination. We end this book as we started it: just the two of us. And yet we must have gone somewhere, because things feel very different now.

Our inability to conceive, we have reminded ourselves in the times when the hope of another month has been extinguished, has come about through no fault of our own. It is neither a failure of ourselves, nor our love, rather some random imbalance in a world where there is already more than enough of that. While we remain childless, we are much closer and stronger for the experience. Our responsibility to each other has grown.

I have reminded myself of this as the years have passed. That we haven't, in fact, been standing still. Weathering the pain of our experience has thickened our skin and strengthened our roots. Like the moorland trees we passed on that midsummer morning, twisted by the elements and bent out of shape by what the world throws at them, but still standing all the same.

Further Reading

Ballard, J.G. 1962. *The Drowned World*. Victor Gollancz, London.

Bisset, Charles. 1762. *An Essay on the Medical Constitution of Great Britain*. A. Millar; D. Wilson, London.

Blom, Philipp. 2019. *Nature's Mutiny: How the Little Ice Age Transformed the West and Shaped the Present*. Picador, London.

Blum, Andrew. 2019. *The Weather Machine: How We See Into the Future*. Penguin Random House, London.

Bohun, Ralph. 1671. *A Discourse concerning the origine and properties of wind*. W. Hall, Oxford.

Burton, Robert. 1621. *The Anatomy of Melancholy*. Vernor, Hood and Sharpe, London.

Chaucer, Geoffrey. 1883. *Parlement of Foules*, edited by T.R. Lounsbury. Ginn, Heath and Co., Boston.

Clare, John. 1827. *The Shepherd's Calendar*. James Duncan for John Taylor, London

Crumley, Jim. 2019. *The Nature of Spring*. Saraband, Salford.

Defoe, Daniel. 1704. *The Storm*. George Sawbridge; J. Nutt, London.

Fitzroy, Robert. 1863. *The Weather Book: A Manual of Practical Meteorology*. Longman, Green, Longman, Roberts & Green London.

Ghosh, Amitav. 2016. *The Great Derangement: Climate Change and the Unthinkable*. University of Chicago Press, Chicago.

Golinski, Jan. 2007. *British Weather and the Climate of Enlightenment*. University of Chicago Press, Chicago.

Groom, Nick. 2013. *The Seasons: A Celebration of the English Year*. Atlantic Books, London.

Harris, Alexandra. 2015. *Weatherland: Writers and Artists under English Skies*. Thames & Hudson, London.

Harrison, Melissa. 2016. *Rain: Four Walks in English Weather*. Faber and Faber, London.

Harte, Jeremy. 1986. *Cuckoo Pounds and Singing Barrows: The Folklore of Ancient Sites in Dorset*. Dorset Natural History & Archaeological Society.

Jefferies, Richard. 1885. *After London*. Cassell and Co, London.

Jennings, Elizabeth. 2012. *The Collected Poems*. Carcanet Press, Manchester.

Langeslag, P.S. 2015. *Seasons in the Literatures of the Medieval North*. D.S. Brewer, Cambridge.

Lewis–Stempel, John. 2015. *Meadowland: The Private Life of an English Field*. Black Swan, London.

Mabey, Richard. 2013. *Turned out nice again: on living with the weather*. Profile Books, London.

Macfarlane, Robert. 2015. *Landmarks*. Hamish Hamilton, London.

Marsham, Robert. 1830. *Indications of Spring observed by R. Marsham at Stratton, Norfolk*. (Read before the Royal Society, 1879.)

Milton, John. 1987. *Paradise Lost*. Penguin, London.

Moore, Peter. 2015. *The Weather Experiment: The Pioneers who Sought to See the Future*. Vintage, London.

Murphy, Patrick. 1834. *The Anatomy of the Seasons, Weather Guidebook and Perpetual Companion to the Almanac*. J. R. Bailliere and Co., London.

Otto, Friederike. 2020. *Angry Weather: Heat Waves, Floods, Storms and the New Science of Climate Change*. Vancouver, Greystone Books.

Parker, Geoffrey. 2017. *Global Crisis: War, Climate Change and Catastrophe in the Seventeenth Century*. Yale University Press, London.

Platt, Edward. 2019. *The Great Flood: Travels Through a Sodden Landscape*. Picador, London.

Pribyl, Kathleen. 2017. *Farming, Famine and Plague: The Impact of Climate in Late Medieval England*. Springer.

Roud, Steve. 2006. *The English Year: A Month-by-month Guide to the Nation's Customs and Festivals, from May Day to Mischief Night*. Penguin, London.

Simpson, Jaqueline and Roud, Steve. 2000. *A Dictionary of English Folklore*. Oxford University Press, Oxford.

Thomas, Dylan. 'Poem In October'. *The Collected Poems of Dylan Thomas: The Centenary Edition*. Weidenfeld & Nicolson, London.

Thomas, Edward. 1914. *In Pursuit of Spring*. Thomas Nelson and Sons, London.

Thomson, James. 1797. *The Seasons*. T. Hepstinall, London.

Viscount Grey of Fallodon. 1927. *The Charm of Birds*. Hodder and Stoughton, London.

White, Rashleigh Holt. 1901. *The Life and Letters of Gilbert White of Selborne*. John Murray, London.

White, T.H. 1958. *The Once and Future King*. William Collins, London.

Wyatt, James. 1872. *A Key to the Seasons and Weather: The Result of Forty-two Years' Observations*. Arthur Hall and Co., Liverpool.

Acknowledgements

My thanks to all those who helped me weather various squalls. Grahame Madge at the Met Office, Paul Stancliffe at the British Trust for Ornithology, Kate Lewthwaite and others at the Woodland Trust, Tim Sparks, Nigel Hand, Alistair McLean at Museums Sheffield, Sue France and Jon Bradley at Green Estate, Ollie Douglas at the English Museum of Rural Life, Kieron Atkinson at Renishaw Hall, Rob Adams at the Spurn Bird Observatory, Kevin Walker and Louise Marsh at the Botanical Society of Britain and Ireland, Viv and Nigel Smith at Claude Smith Ltd, Al Dix, Anne Storm and Norman Ackroyd.

My thanks also to all the readers of my Weather Watch column who, as well as being quick to correct my mistakes, have sent me so many entertaining and moving letters over the years. They make writing a privilege.

Once again my ace Bloomsbury editor Alice Ward has been invaluable in helping steer me through the fog. I am grateful to my copy editor Mari Roberts for her thoughtful suggestions and Jim Martin for taking a punt on me many seasons ago. Thank you also to my agent, Antony Harwood, and Jon Day for advice on early drafts of the book.

I am also indebted to my Telegraph editors, especially Vicki Harper, Jessamy Calkin and Jane Bruton for all their support.

Finally, thanks to my friends and family for their kindness and always keeping my outlook bright. To my mum, for her sharp eye for grammatical errors and being a sounding board for so much else. And most of all to my wife, Liz, for trusting me to tell our story in the hope it might help others.

Index